Structural Engineering

Also available from E & FN Spon

Aluminium Construction
J.B. Dwight

Concrete in the Service of Mankind
Edited by R.K. Dhir

Containment Structures: Risk, safety and reliability
Edited by B. Simpson

Design Aids for Eurocode 2: Design of concrete structures
Edited by the Concrete Societies of the UK, The Netherlands and Germany

Design of Masonry Structures
Edited by A.W. Hendry, B.P. Sinha and S.R. Davies

Design of Structural Elements
C. Arya

Durability of Concrete in Cold Climates
M. Pigeon and R. Pleau

High Performance Concrete
P-C. Aitcin

High Performance Fiber Reinforced Cement Composites 2
Edited by A.E. Naaman and H.W. Reinhardt

Reinforced Concrete Designer's Handbook
C.E. Reynolds and J. Steedman

Reinforced and Prestressed Concrete
F.K. Kong and R.H. Evans

Structural Analysis: A unified classical and matrix approach
A. Ghali and A.M. Neville

Structural Assessment: The role of large and full-scale testing
Edited by K.S. Armer, J.L. Clarke, G.S.T. Armer and F.K. Garas

Structural Design of Polymer Composites
Edited by J.L. Clarke

Structural Dynamics
M. Paz

Structural Mechanics: A unified approach
A. Carpinteri

For more information about these and other titles please contact:
The Marketing Department, E & FN Spon, 2-6 Boundary Row, London, SE1 8HN Tel: 0171 865 0066

Structural Engineering: History and development

Edited by

R.J.W. Milne

CRC Press
Taylor & Francis Group
Boca Raton London New York

CRC Press is an imprint of the
Taylor & Francis Group, an **informa** business
A TAYLOR & FRANCIS BOOK

CRC Press
Taylor & Francis Group
6000 Broken Sound Parkway NW, Suite 300
Boca Raton, FL 33487-2742

First issued in paperback 2019

© 1997 by Taylor & Francis Group, LLC
CRC Press is an imprint of Taylor & Francis Group, an Informa business

No claim to original U.S. Government works

Typeset in 10/12pt Meridien by Best-set Typesetter Ltd., Hong Kong

ISBN-13: 978-0-419-20170-0 (hbk)
ISBN-13: 978-0-367-86475-0 (pbk)

A catalogue record for this book is available from the British Library

Visit the Taylor & Francis Web site at
http://www.taylorandfrancis.com

and the CRC Press Web site at
http://www.crcpress.com

Contents

Foreword

One of the highlights of my year as President of the Institution was the celebration of the 60th anniversary of the grant of our Royal Charter. A number of events were held to celebrate. This publication provides a record of our seminar on the history and development of structural engineering held on 'Jubilee Day' – May 4th 1994.

Eight Institution gold medallists participated in the seminar, giving papers from their own experiences covering broad aspects of the history and development of structural engineering from philosophy, through research, analysis and design.

During the day of the seminar I was struck by the great spirit of community between one of the most distinguished groups of members that we have in the Institution, both between each other and with all of the many members who attended the seminar, of whatever age and specialism.

The papers in this publication provide a record of the events of the day and have many points of interest, both in establishing an historical context for current practice and in giving pointers for the way ahead.

Nothing can replace the spirit and interest of the day, but I hope that the reader will get a flavour of what was one of the most successful events of the Jubilee celebrations.

Howard Taylor
President 1993–1994

Institution medallists

Gold Medal

For personal contributions to the advancement of structural engineering

Henry Adams	1922
John Fleetwood Baker	1953
Eugene Freyssinet	1957
Hardy Cross	1958
Felix Candela	1960
William Henry Glanville	1962
John Guthrie Brown	1964
Pier Luigi Nervi	1967
Alfred Grenvile Pugsley	1968
Knud Winstrup Johansen	1971
Yves Guyon	1972
Ove Nyquist Arup	1973
Henry Charles Husband	1973
Fritz Leonhardt	1975
Oleg Alexander Kerensky	1977
Nathan Mortimore Newmark	1979
Riccardo Morandi	1980
Alex Westley Skempton	1981
Alan James Harris	1984
Frank Newby	1985
Michael Rex Horne	1986
Alan Davenport	1987
Anthony Ray Flint	1988
Gerhard Jacob Zunz	1988
Jorg Schlaich	1990
Edmund Happold	1991
Olgierd Cecil Zienkiewicz	1991
Santiago Calatrava	1992
Anthony Hunt	1994

1
The Institution of Structural Engineers: then and now

Cyril Morgan

Although we are celebrating the 60th anniversary of the Charter being granted to the Institution, it should be remembered that the Institution started much earlier, with the formation of the Concrete Institute and the first meeting of its Council in the Smoking Room of the Ritz Hotel in July 1908. The years around the turn of the century were significant for the introduction of new materials and technologies into general UK practice, particularly reinforced concrete and steel frame buildings. It was entirely appropriate that a specialist body should be formed to meet the needs of those interested and involved in using these new methods and not surprising that the professional exchange should develop into a body for structural engineering as a whole. This came in 1922 when the Council of the Institute resolved to change the name to the Institution of Structural Engineers, with the first Council and Officers of the new-named body taking office in 1923 when the total membership was 1330.

Even before the change of name, some of the characteristics of the present Institution were already in place. The library had been founded and in 1922 a monthly journal was started which, in 1924, took the title *The Structural Engineer*. In retrospect, this was a major step forward. *The Structural Engineer* is now recognized as being one of the foremost journals to deal with the theory and practice of structural engineering and in its combination with *Structural news* also provides a link for news, views and Institutional business across the whole of the membership.

The years before the granting of the Charter were all of steady development. Branches of the Institution were formed. First, Lancashire & Cheshire in 1922 and then, in quick succes-

Structural Engineering: History and development, Edited by R.J.W. Milne. Published in 1997 by E & FN Spon, London. ISBN 0 419 20170 X.

sion, Western Counties, Yorkshire, Midland Counties, and the Wales and Scottish Branches.

Examinations for entry had been introduced in 1920 but, with membership reaching 3000 in the early 1930s, the qualifying examinations were revised to include a compulsory paper – structural engineering design and practice. It was with this background that the Institution successfully petitioned for a Royal Charter which was granted in perpetuity by His Majesty King George V in May 1934.

Following the grant of its Charter, the Institution continued to develop apace. The first overseas branch was formed in South Africa in 1936, and the Institution became more active in the production of technical memoranda and writing Codes of Practice – an activity later taken over by British Standards Institute (BSI), but now mirrored by the series of Manuals prepared by the Institution in association with the Institution of Civil Engineers.

Remarkably, during the period from the granting of the Charter to the present day, the Institution's administration has been in the hands of just three long-serving Secretaries. Major Reginald Maitland, OBE, was Secretary from 1930 to 1961 and saw the Institution through the stages of the Charter, the technical developments of the 1930s, and the years of the Second World War. He was followed by Cyril Morgan, OBE, who was Secretary until 1982 when he was succeeded by Derek Clark who retires at the end of June in 1994.

Cyril Morgan's period as Secretary was marked with growing status for the Institution itself and greater consciousness of links with other professional bodies. A key event in this was the gas explosion on 16 May 1968 in a 23-storey building, causing a disastrous cumulative collapse of a whole corner segment of a block of flats in Plaistow, East London, which (for ever after) is known simply as 'Ronan Point'. Cyril Morgan acted as Secretary to the inquiry, and the Institution was centrally involved in providing advice to Government and recommendations on future design practice related to avoiding progressive collapse. This event and the efforts of members and those closely involved provided Government, local authorities and the public at large with a new awareness of the practice of structural engineering and the role of the Institution. Since that time, the Institution has continued to be consulted regularly on Building Regulation matters.

From 1961 onwards we had the beginnings of the present unification debate, with the Institution involved in discussions which led to the formation of the Council of Engineering Institutions (CEI) under Royal Charter in 1965. Eventually, this led to our use of the CEI's Part 1 and Part 2 examinations in place of the Institution's Part 1 and Part 2, but with retention of our Part 3 which, by then, had become widely recognized both in the UK and abroad as a definitive statement appropriate to the practice of structural engineering. The CEI was followed by the Finniston Inquiry and the formation of the Engineering Council on the new 'engine of change' for the professions. *Plus ça change, plus c'est la même chose* – the quest for a unified general system coupled with sufficient independence for the specialist institutions to serve their own members and respond to the requirements of their Charters continues to the present day and has been reactivated by the present Chairman of the Engineering Council, Sir John Fairclough.

Cyril Morgan retired as Secretary in October 1982 from an Institution with high prestige and a total membership of around 15 000. He was succeeded by Derek Clark who has seen the Institution through a period of intense development and high activity in the period since the 50th anniversary celebrations in 1984.

During this period the membership has grown to some 23 000, stimulated by greater attention to our grassroots and particularly the adoption of a free membership scheme for students attending courses approved by the Joint Board of Moderators. The membership continues to increase, borne by the value placed on the Institution qualification, the progression from Student to Graduate to Member, and the development of overseas divisions and Joint Membership Agreements based on the Institution Part 3 examination. This has underpinned the shift of the Institution from a purely UK body to one that is truly worldwide, but with its centre in London. With the formation of joint divisions in Hong Kong, South Africa, Singapore and Islands of the West Indies, and with others in train, this process will continue both on the basis of the qualification but also as a means to facilitate exchange of technical and professional expertise through *The Structural Engineer* and learned-society activities.

The period has also been notable for the development of a robust programme of continuing professional development in the Institution, enhancement of the support for professional activities with the appointment of an Assistant Director of Professional Affairs, and an enlargement of the technical secretariat. This has led to new major events, including the series of

Kerensky International Conferences (with the 1994 conference to be held for the first time outside the UK, in Singapore), and a wider range of technical publications dealing with such matters as sports grounds, building inspections, and subsidence.

Besides the growth in technical and professional support, the administration of the Institution has been modernised with computer systems, wordprocessing, and desktop publishing. The building itself has been completely refurbished and now provides an appropriate base for a leading Institution close to Government and available to its own members. Because of the range of activities and services being provided it was considered desirable to establish a trading company to deal with all Institution-linked activities that did not fit the requirements of its charitable status. This has enabled further new initiatives to be undertaken – a *Directory of Firms*, promotional leaflets for companies, servicing of other bodies, etc. – which provide a surplus that is covenanted back to the Institution.

On the wider strategic front, the Institution, under the presidency of Professor (subsequently Sir) Edmund Happold, was instrumental in 1987 in forming, with other bodies, the Building Industry Council which now, as the Construction Industry Council, has developed strong and direct communication with Government and increasing respect and commitment from the industry. The Institution continues to play a key independent role in the discussions on the desirable 'new relationship' between the professions and Engineering Council, particularly in emphasizing the role, value and necessary independence of the specialist institutions.

The last 10 years have been ones of considerable growth, change and response to outside influence. The later years have coincided with a deep recession in the construction industry, followed by a period of minimal (or at best slow) recovery. It is not surprising, therefore, that the Institution is now experiencing a few small growing pains shown in the balance sheet and stretched staff resources. This is a temporary situation; the underlying strength of the Institution in its growing membership, qualification and sense of purpose cannot be doubted. The Institution will continue to develop to meet the needs of the next century, as it has so successfully done through the 1900s. Besides being our anniversary, 1994 is also another year of change with the appointment as Chief Executive on 1 January of Dr John Dougill, who additionally takes over as Secretary in July. We look forward to a period of continuing expansion of our influence in UK, Europe and abroad and a maintenance of our status as the premier Institution for structural engineering worldwide. The future beckons.

Note

This Chapter is based partly on the article 'A history of the Institution of Structural Engineers 1908–1983' by Cyril Morgan, Secretary of the Institution from 1961 to 1982, *Structural Engineering – Two centuries of British achievement* (Tarot Print Ltd, 1983, pp. 187–188) and notes on subsequent developments provided by Derek Clark, Secretary 1982–1994.

It is a matter of sadness that Cyril Morgan died, after 11 weeks of intensive illness, on 19 April 1994.

aul Swiridoff

2
Reflections on 60 years of structural development

Fritz Leonhardt

Fritz Leonhardt has spent much of his career in Stuttgart University, graduating from there in 1931, working as a professor from 1958 to 1974 and spending two years in the 1960s as the rector of the university. This has paralleled his career as a senior partner in Leonhardt, Andrae und Partner Consulting Engineers.

As an engineer he has concentrated on the development of novel systems of bridge engineering and on the design of other complex civil engineering structures, primarily in prestressed and reinforced concrete. He has also worked to introduce new construction methods for bridges. He has been responsible for a number of suspension or cable-stayed bridges across the Rhine, notably the slender hollow-box beam large-span bridge at Cologne-Deutz, and several TV towers and large span structures.

It is the 60th anniversary of this honourable Institution, which was founded in 1934. I had graduated from the University of Stuttgart three years earlier, in 1931. Thus, I can cover these 60 years with my professional experience of 63 years, beginning as a bridge engineer for the German *Autobahnen*.

What does the term structural development mean? It means man-made – made by engineers, who design and build structures.

Therefore, I first wish to look at these engineers – how must they be trained to become creative, to find innovative solutions which overcome traditional work and cause development?

Looking for sources of development I may mention six influences:

1 Inventions based on actually new ideas – rather rare.
2 Failures of or damages to structures which challenge improvements.

Structural Engineering: History and development, Edited by R.J.W. Milne. Published in 1997 by E & FN Spon, London. ISBN 0 419 20170 X.

3 Research and testing of materials and structural elements.
4 Decrease in the costs of structures for economical reasons.
5 Competition in design and construction.
6 Contributions of mechanical engineers developing better or new construction equipment or construction methods.

How are engineers created? They must have a solid basic knowledge of the many sciences needed for the profession. These are mathematics, normal physics and chemistry, mechanics, statics, some dynamics, strength and behaviour of building materials, building physics, soil mechanics and, in our days, informatics. They must also be acquainted with modern construction methods and the available equipment.

Our engineers should know the structural qualities of structural systems like the beam, slab or plate, frame, arch, suspensions, folded plates, shells, trusses, space trusses and even networks and membranes.

It is most valuable if they can see demonstrations of the behaviour of these structural systems by model tests, to see the deformations under loads – to learn what stability means etc. They should have carried out or seen tests on structural members up to failure.

This is a great amount of knowledge which must be learned and digested. But I stress that basic knowledge in these fields will be sufficient – in each field there are specialists with sophisticated knowledge. Thus, practising engineers must only know that such experts exist and may be contacted if they encounter a problem for which they do not feel sufficiently competent. This means that in our days, the engineer must principally be educated to team-work and to interdisciplinary thinking.

During their training, engineers should travel and visit interesting structures or construction sites and consciously see and analyse such structures. They should not hesitate to ask questions of the field engineers or craftsmen on such occasions.

This background of learning is most important because design abilities improve with the quantity and quality of structures which the young engineer has seen, studied and registered mentally. The engineer must have a rich repertoire of design possibilities in mind to draw from when starting a design.

The state of the art of our engineering sciences has also developed during these 60 years and I wish to give you an interesting example of how this has influenced structural development.

In the 1930s statically determinate structures were preferred, especially in cases of poor foundation conditions with settlement to be expected. Single span beams were quite common. The methods to calculate the forces in statically indeterminate structures were not well known and were tedious to be done by hand. This was the situation in which I found myself when I had to design my first bridges. But obviously, continuous beams or frames had considerable advantages for service requirements and also for safety against failure. It took a few years to get continuous beams, frames and fixed arches accepted. As a consequence, expansion joints had to be developed for large movements, and even watertight joints were needed. Bearings were designed to allow adjustment to different settlement. However, after some years we built with long continuous beams, even in a region where considerable mining subsidence was expected.

As the German railways started around 1980 to build bridges for the new high speed trains, there was a plan to build only 40m single span beams in order to avoid rail expansion joints. This would have been a disaster for high viaducts. In long discussions I succeeded to convince them that, especially for high speed, continuity is a necessity. We have subsequently built 1200m long continuous beams for these high speed trains.

In the computer age, numerical analysing of structures is, of course, no longer a handicap in the choice even of complicated structures. In some cases it is not so much the calculation of static forces, but the calculation of the geometry of a construction. This was the case for the networks of the Olympic Stadium in Munich (1970), for which Professor J. Argyris provided a FE program to calculate the exact length of the thousands of strands and ropes just in time.

Let me now give an example where the need to economize gave the impetus to developments.

When we started to build *Autobahn* bridges around 1934, the decks of steel bridges were rather heavy, weighing between 1000 and 1200kg/m². I proposed plane steel plates stiffened with ribs with only about 7cm of asphalt wearing surface, bringing the weight down to about 200kg/m². Test runs had been made to find devices for reliably fixing the thin asphalt layer to the steel plate, even under high and low temperatures. A small test bridge was built (Fig. 2.1). But the construction

Figure 2.1 The first small test bridge with an orthotropic steel deck (1934).

company complained of problems with welding deformations, and this stopped the development of these light weight steel bridges. However, after the Second World War construction companies themselves came back proposing such steel decks. During the war automatic welding was developed, which kept welding deformations under control. Such bridge decks are now known under the scientifically sounding name of 'orthotropic plate' (a contraction of 'orthogonal anisotropic stiffened plate').

Welding of steel bridges was strongly promoted around 1934–38 by our construction firms, and this resulted in long span, continuous plate girder bridges. However, there came a bad set back. On a very cold winter night in 1937, a flange 40/1000mm² in cross section on a bridge near Berlin broke by brittle rupture. Fortunately, the bridge did not fall down, because it was continuous, but this incident started an enormous amount of research work to develop steels suitable for welding and to find welding techniques that would reduce welding stresses or even anneal them.

It was mainly Professor Klöppel of Darmstadt who carried out and published these tests, which also had an influence on the welding details of the orthotropic steel decks of our bridges.

If British engineers had studied these test results, they could have avoided the problem of cracking in the orthotropic deck of the Severn Bridge.

Another example of the influence of testing on structural details may be mentioned. In 1938 testing with pulsators allowing millions of load cycles in a reasonable time were just coming into use. I had to build the suspension bridge across the Rhine in Cologne-Rodenkirchen (Fig. 2.2). I was sceptical about the fatigue strength of the anchorage of the ropes in zinc filled sockets as I saw that this was done at temperatures above 450°C. This must damage the strength of the cold drawn wires. The test result was shocking: we got only 80 to 100N/mm² allowable stress amplitude with wires of 1600N/mm² strength. Consequently, we developed together with the Swiss engineering company BBR, Zürich a new type of anchorage with a cold filling, the so called high amplitude (HIAM) anchorage. We obtained allowable fatigue stress amplitudes of 200 to 250N/mm². Such a high fatigue strength is especially useful for cable stayed bridges, in which the back stays can undergo considerable stress changes. Of course, at these anchor sockets, bending stresses due to oscillations of the ropes or cables have to be strictly avoided if such high stress changes are expected.

Our designs are ruled by codes which specify certain limits, and these are often a handicap for structural development. Immediately after the Second World War I had a chance to ignore such code limits and achieved a breakthrough. This was in 1946, when I was working on the design of the

Figure 2.2 Bridge across the Rhine in Cologne-Rodenkirchen; the first suspension bridge with a continuous plate-stiffening girder (1938).

Cologne-Deutz Bridge across the Rhine (Fig. 2.3). The old bridge was a self-anchored suspension bridge with a main span of 184m. I had the ambition to build a deck bridge without cable-suspension on the existing piers, but the available depth enforced a slenderness of the continuous beam (1/56 in the span, 1/24 at the piers) which had never been dared before and which gave deflections under live loads of L/260 – the code's limit was L/700. But in 1946 no German control was yet available and the slender bridge was built. For me it was clear that this code restriction was wrong. The serviceability of such bridges depends not on the relative deflection, but on limits of curvature and inclination of the deflection line and on oscillations. The code was soon changed. This Rhine bridge was the first very slender steel box girder bridge and it opened the field for such bridges.

Let me now come to inventions. In 1941 Eugène Freyssinet

Figure 2.3 This bridge across the Rhine in Cologne-Deutz was the first slender box girder bridge (1946).

wrote his famous paper *Une révolution dans l'art de bâtir* [1] and, in fact, his invention of modern prestressed concrete caused a revolution. It was based on an early patent saying that the steel had to be stressed to at least $400N/mm^2$ in order to minimize losses of steel stresses due to shrinkage and creep of the concrete.

The French mainly built structures with T-beams with very thin webs, like Freyssinet's famous Marne Bridge in Luzancy and Esbly. These T-beams were mostly prefabricated. I had box girders in mind and carried out large scale tests using them, studying bonded and unbonded tendons. The results convinced the German railway bridge engineers and thus we could build in 1951 the first German prestressed concrete railway bridge in Heilbronn – a five span, skew multi-cell continuous plate with concentrated prestressing cables – continuous over 107m length. The bridge is still in good condition.

Box girders were my favourites. In 1950 I built a three span

Figure 2.4 The 96m long bridge across the Neckarkanal in Heilbronn (1950) was the first long span prestressed concrete bridge.

continuous girder with a 96m main span, again with an extreme slenderness, cast *in situ* (Fig. 2.4). In 1951 Ulrich Finsterwalder – my main competitor – built the first prestressed concrete bridge across the Rhine at Worms using the free cantilevering method, also with a single cell box girder.

In 1953 I had to design a curved and skew bridge over the Danube at Möhringen. On this occasion we realized the advantages of these continuous box girders for curved bridges, supporting them with single bearings or slender columns in the axis. This bridge became the prototype for the many curved bridges at *Autobahn* junctions and for elevated streets in cities.

The French turned also to box girders, but developed construction with precast elements with match-cast joints. Damage at these joints due to differential shrinkage and creep could have been foreseen, and it did indeed occur. This made me develop the incremental launching method, where short portions are fabricated step by step at the abutment and are then launched over the spans (Fig. 2.5). This allows reinforcement through the joints and keeps the temperatures due to hydration etc. under control. Consequently we got perfectly tight joints with a ductile behaviour for ultimate safety. This incremental launching method is now widely used in Europe, even in France, where it is used mainly for new railway bridges.

This development was again accompanied by a long series of

Figure 2.5 Incremental launching to build long, continuous, prestressed concrete bridges – even curved bridges – began in 1962.

tests to get reliable and simple movable bearings – the rubber pot bearing sliding on Teflon. These bearings were soon common all over the world under various names.

Let me continue with inventions as a source of structural development. I wish to refer to those creative engineers who designed shell and membrane structures, for example Ulrich Finsterwalder, Eduardo Torroja, Felix Candela and especially Heinz Isler, with his interesting methods for finding the optimal shape by experiment. Most of these shells are aesthetically very pleasing and it is a pity that the expense and difficulty of their construction has dissuaded many engineers from building shell structures.

In this group I must also mention Frei Otto, with his Institute for light weight structures at the University of Stuttgart. His creative spirits have offered a large number of new structural systems, mainly tent shaped tension structures or membranes. The stadium and the sports hall for the 1972 Olympic Games in Munich were based on his ideas. His seminars received wide international acclaim. The development of his structures was again accompanied by a large number of tests.

Let me now come to the impetus which structural failure can give to the development of new structures – the collapse of the Tacoma suspension bridge in 1940 due to wind oscillations has had a strong influence on my personal work.

My idea was to shape the cross section of the bridge deck in a way to reduce the wind forces and to avoid wind eddies, which cause bending and torsional oscillations (Fig. 2.6). The bridge deck should be as shallow as possible with aerodynamically shaped noses. By using only one cable and inclined

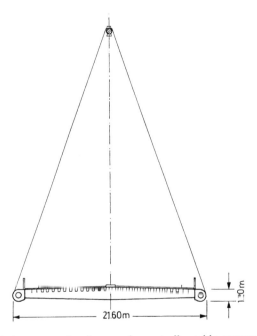

Figure 2.6 A cross section for aerodynamically stable suspension bridges tested in the NPL, Teddington.

hangers, torsional oscillations could be avoided. This type of monocable suspension bridge was patented in 1953.

Wind tunnel tests would have proved the aerodynamic safety of this new type, but at that time we had no wind tunnels available in Germany and so I went to Mr Scruton at the National Physical Laboratory (NPL) in Teddington, who at that time was conducting wind tunnel tests for the Severn Bridge with the usual stiffening trusses. The result of tests on a flat deck were so convincing that we could offer this bridge type for the Tejo Bridge in Lisboa, which has a main span of 1100m. By political influence, however, an American type suspension bridge with large stiffening trusses was built there. Meanwhile, Freeman Fox changed their design for the Severn Bridge to a flat box section with thin wind noses.

The Americans and later also the Japanese kept their stiffening trusses, increasing their depth to 10m (Tacoma Bridge) and even to 12m for the Mackinac Straits Bridge – I used to call them 'old fashioned American type suspension bridges'.

My next chance to build an aerodynamically shaped suspension bridge was in the competition for the Rhine Bridge at Emmerich in 1961, which has a main span of only 500m (Fig. 2.7). In order to prove the reliability and especially the truss effect obtained by inclined hangers, we had made an 18m long

Figure 2.7 A monocable-suspension bridge for crossing the Rhine in Emmerich, with a span of 500m and the cross section shown in Fig. 2.6, was designed in 1961, but never built.

model. The construction firms offered this design about 15% cheaper than the old fashioned design with stiffening trusses. Unfortunately, the bridge engineer of our Federal Ministry of Transport refused to accept this design – he considered it too progressive, too far beyond established experience. This was not the only case where new developments were handicapped by the lack of willingness to share responsibility by our German officials.

Meanwhile, the Severn Bridge was built. In Denmark, the Lillebelt Bridge got a slender box girder, its wind stability confirmed by Selberg's wind tunnel tests in Trondheim. Soon the Humber Bridge followed.

When designing the bridge crossing Burrard Inlet in Vancouver, I had more wind tunnel tests carried out at the NPL in Ottawa by Mr Wardlaw to optimize the shape of the wind nose. We also proved that the bottom of the deck structure can remain open: it need not be a closed box.

Meanwhile, the aerodynamically shaped flat box section was being used for modern suspension bridges, for example the Bosporus Bridges and the Great Belt Bridge, with a 1900m main span. For the Messina Bridge, a 3300m main span is proposed, but there is in my opinion a limit of constructability which may count here. It would be very difficult or even impossible to compose the 1.5m diameter cables on a cat walk 80–400m above the sea which cannot be stiffened against wind oscillations.

At the same time as this development of modern suspension bridges, cable stayed bridges started to prove their technical and economic superiority over suspension bridges. The basic impetus came from Dischinger, who said that cable stayed bridges have superior stiffness if the cables are highly stressed (up to $900N/mm^2$), using cables of the high strength steel developed for prestressing concrete structures.

Our development started with the designs for the famous Düsseldorf Bridge family across the river Rhine from 1952 to 1956 (Fig. 2.8). We collected construction experience using the inclined cables for free cantilevering. They were spaced more than 30m apart, making auxiliary equipment necessary. In 1964, Homberg built his Rhine Bridge in Bonn, with very closely spaced cables (2.26m) which allowed a free cantilevering construction without auxiliary means. Simultaneously, the anchorages of the cables were simplified. The multi-cable stayed bridge was born.

New knowledge was again obtained by tests. As we were

Figure 2.8 Kneibrücke Düsseldorf – one of the three cable stayed bridges of the Düsseldorf 'Bridge Family' designed in 1952–6.

working on the Paraná Bridge at Zarate-Brazzo Largo in Argentina, for a highway and a railroad carrying 400m long freight trains, we came to realize that aerodynamic shaping of the cross section was useless for wind stability with such trains on the bridge. We carried out dynamic model tests at the Ismes Institute in Bergamo which proved that the damping of these multi-cable systems is so strong that there is no danger of resonance oscillations for cross sections without a wind nose. So we built the bridges without a wind nose and they continue to behave well.

For cable stayed bridges with concrete decks, for example the one I designed in 1971 for the Pasco-Kennewick Bridge over the Columbia River in the USA, we wished to get away from triangular box sections at the edges and to choose only flat slabs with very shallow edge beams (Fig. 2.9). The question was – is there any danger that such slender decks could buckle under the high normal compression forces of these large suspended cantilevers? René Walther of the Ecole Polytechnique de Lausanne conducted a large test with a very

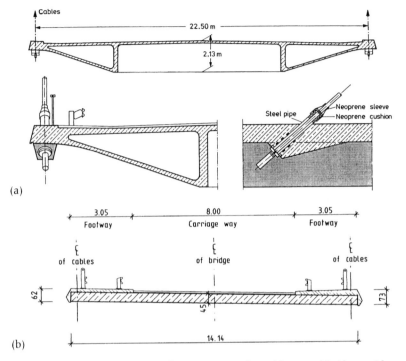

Figure 2.9 Development of cross sections for cable stayed bridges with a concrete deck: (a) Pasco-Kennewick Bridge over the Columbia River (span 300m, 1972); (b) Evripos Bridge in Greece, (span 200m, designed by J. Schlaich, 1988).

thin deck slab without an edge beam. The test proved that in spite of considerable deflections, there was no buckling. This test opened the way to cable stayed concrete bridges with only a deck slab without edge beams or with very slender edge beams, such as those built by Walther across the Rhine in Diepoldsau, by Schlaich in Evripos in Greece and by Svensson of Leonhardt, Andrä and Partner for the Helgeland Bridge in Norway.

Thus, frequent testing has helped to drive forward the development of these cable stayed bridges. They grow like mushrooms. More than 120 have been built during the last 50 years, of which more than 30 are in China.

In 1967 I had proposed a cable stayed bridge with a span of 1800m across the Straits of Messina for a railroad and highway. At present, spans above 800m are under construction. The Normandy Bridge in France has an 856m main span, the

Tatara Bridge in Japan will be 890m. In time development will go beyond these spans.

As an impetus to development I mentioned also the contributions of mechanical engineers developing new construction equipment. This is a wide but important field. Just consider the impressive floating cranes for moving and placing enormously large bridge portions with weights of several thousand tons. Japanese steel constructors were the first to use such giants. For the western part of the Great Belt Bridge in Denmark 6000t, 108m long box girders were placed by a catamaran crane between the pier heads.

Also in the field of foundations, new construction methods have been developed, mainly in Japan. Daring engineers have developed offshore structures for oil extraction. Such outlooks show the very wide field of our profession – serving mankind in so many ways.

I hope that you will forgive my treatment of 'reflections on 60 years of structural development' mainly by examples of my personal experience and contributions.

New tasks now face civil engineers, with the wide and complex problems of ecology to keep our earth inhabitable for its increasing population. The misuse of limited resources must be stopped – for instance, the change to solar energy must come. Thus, many challenging tasks stand ahead of us for creative engineers who are able to develop new solutions. This offers a promising future to our profession.

Reference

[1] Freyssinet, E. (1941) Une révolution dans l'art de bâtir. Les constructions précontraintes, *Travaux*, November.

3
Philosophy of design with particular respect to buildings

Edmund Happold

Sir Edmund (Ted) Happold was born in Leeds, and remained a gritty Yorkshireman all his life. He began his engineering career as a site engineer for Sir Robert McAlpine & Sons. In 1957 he joined Ove Arup, before going to the USA to work for Severud, Elstad and Krueger from 1959–61. Then he returned to Britain and Ove Arup and Partners, where he was involved in the design of Coventry Cathedral, the Sydney Opera House and the Pompidou Centre. Perhaps the achievement which summed up his greatest design skills was the light, flowing, asymmetrical grid shell across the top of the garden exhibition at Mannheim.

In 1976 his career changed with a move to Bath to set up Buro Happold. At the same time he became Professor of Building Engineering at the university there, and worked to combine the teaching of building engineering and architecture. He also co-founded and chaired the Construction Industry Council. A life of brisk activity took its toll on his heart (as he said 'The structure's fine, it's the pump's the problem') and he died in 1996 at the age of 65.

Philosophy, if the Oxford dictionary is to be believed, 'is the pursuit of wisdom or knowledge of things and their causes, whether theoretical or practical'. Historically there were three branches – natural, moral or metaphysical. In simple terms philosophy is the system which a person forms for the conduct of life [1].

I suppose I brought the task of contemplating this subject on myself by a talk I gave at the Institution in February 1992 when they gave me a Gold Medal and which I called 'The Nature of Structural Engineering' [2]. I thought, and still think, that such a discussion is important because engineers

Structural Engineering: History and development, Edited by R.J.W. Milne. Published in 1997 by E & FN Spon, London. ISBN 0 419 20170 X.

tend to remain anonymous and we are actually all rewarded nowadays not by personal recognition but by how the public sees the significance of engineering or how its cultural meaning is extended. As Pacey says [3], while people recognize our creativity and daring, few engineers indeed are recognized as benefactors of mankind.

I would not wish to repeat what I wrote in the previous paper, the early part of which leaned on a collection put together by Mitcham and Mackay [4] and in which they divided questions into three orders. The first order being about technical and practical competence – such as quantifying loadings in different situations, defining allowable stresses in materials, analysing structural behaviour, sizing elements and so on. Not my brief today – except if we do not extend our breadth of competence, the days for this Institution are numbered because society needs fewer and fewer traditional structural engineers.

But then that is a good example of a second order question, one concerned with the nature and breadth of our discipline. Is engineering merely applied science? What is the meaning of technical efficiency and how does it differ from economic efficiency? What is the relation between engineering and architecture? – that is another way of putting the theme of today. They are fundamental questions because they define our discipline – should determine our educational training programmes – and should reflect society's needs.

Which leads me to third order questions which relate to the social, economic, political and moral problems which are caused by engineering activity – those which underpin the conduct of your life.

The second order questions depend on understanding the first order, third order questions cannot be approached before having some knowledge of the second order. They include history and futurology – the one has clues for the other. But as human perception and reaction have changed very little over several thousands of years, they are usually linked. However, technology has advanced and largely because of it our organizational methods and social behaviour have changed.

I tried in the previous paper to explain the nature of engineering activity, its relation to science, its objectives – an aim for efficiency, different levels of design and so on. The whole to give a definition that the purpose of engineering design is to organize the construction and operation of some artefact which amends the physical world around us in order to meet some recognized need.

It sounds like building – or constructing any other structure. But the whole is not quite as easy as that.

Building, or at least shelter, is as old as man. The cave to the henge and so on for fixed habitation, the tent of skins or fabrics as a movable one – climatic moderators.

Not that their manufacture represented a designed product. All artifacts, and the tools to make them, probably came about because a primitive man or woman was presented one day with such an arrangement as to be able to see a future implement in it – the discovery of tools, learning by trying and adapting. Making by craftsmanship. As passing on this knowledge by apprenticeship tended to be limited to traditional knowledge, it was difficult to develop invention.

History tells us that (forgetting military engineering) thinking out what you want to build before you start building (because it is cheaper and less harrowing to do so) really first lay with architects. It expresses, I suspect, an essential belief in mankind, that there should be more than just utility in a product – or an action – in life, if you like. The stone age hunters took time out to paint cave walls. They carved the cave entrances. The largest early structures were henges, medieval ones were mosques and cathedrals and so on. More than utility had to be provided in the product, a further ingredient was needed, which we call beauty, delight or such like that appeals to the senses, art if you like.

You reflect your wealth or your rebellion by providing more art.

So architecture came first and to reinforce that I can probably do no better than to list for you the chapter headings from Steen Eiler Rasmussen's *'Experiencing Architecture'* [5]:

- Basic observations
- Solids and cavities in architecture
- Contrasting effects of solids and cavities
- Architecture experienced as color panes
- Scale and proportion
- Rhythm in architecture
- Textural effects
- Daylight in architecture
- Color in architecture
- Hearing architecture

The book never mentions technology. It is about the senses. It is about taste. It tells engineers like us what we should never forget – the primacy of architecture in building. Because architecture is about satisfying people. It is about planning, surfaces, forms and spaces – to satisfy people.

Engineering as we know it today is much younger than architecture. It was a new way of thinking – what Kuhn called a 'paradigm shift' [6] – that developed modern technology. It came with the scientific revolution, the technology of making the equipment needed to be able to observe and analyse science. The two are irrevocably linked, the engineer and the physicist break their search down into the same processes, observing nature, analysis, experimenting, though the physicist is collecting knowledge and technology is about developing skills, knowing what and knowing how. So engineering is about satisfying nature. Perhaps the best description was given by Thomas Tredgold in 1828 when supporting the founding of the Institution of Civil Engineers he wrote 'the profession of civil engineering being the art of directing the great sources of power in nature for the use and convenience of man'.

Civil engineering constructs a new nature, a super nature, between mankind and original nature. Which is what architecture does not do. So to restore our egos we can comfort ourselves with two arguments:

- Architects can do little without us since we provide the possibilities they use – be it structure, services, construction and operational methods and so on.
- At one end, it is engineering art, making and doing something which extends the vision of those who are watching, which is used to effect change in building. Architectural visual change is usually rooted in engineering invention.

All architects depend on our engineering services. Some architects pick up the visual imagery of engineering like magpies, to embellish, sometimes mindlessly, their buildings. Some change the engineers they work with in order to both extend the supply of solutions and to ensure their design reputations are theirs alone.

There is nothing wrong with any of this. But, as Kipling says, we are Martha's sons. Certainly aggressive attitudes have declined recently as mutual respect has risen. And history often, will out.

Galbraith has written [7] 'what is common to most successful technical enterprises is the inevitability of collective decision making and guidance in which specialists participate, contributing the needed knowledge or expertise.' That decision making is dependent on dialogue, in turn dependent on enough shared language; understanding not just the concepts but the intentions.

If one thinks about this it is incredibly difficult. Even if the disciplines have the same common everyday language the architects' design techniques are primarily visual, the engineers' are usually either experimenting physically or analysing numerically. Further to that, there is often a deliberate separation of education promoting differing professional values and mystification. An example of this is that, with the exception of a very few practitioners, those engineers most admired by architects are not the same as those most admired by engineers and *vice versa*.

This point is a very interesting one because some philosophers, such as Bunge [8], have argued that technology appears when pre-scientific crafts are replaced by technological theories based on scientific laws which explain their effectiveness. All action comes under one of two areas of theory. There are substantive technological theories which are straightforward applications of pre-existing scientific theories and there are operative technological theories which are directly concerned with 'the operations of men and men machine complexes' – scientific management. Faced by the architects' belief – and success – in 'intuitive' design one may have a preference for philosophies such as Mumford.

Mumford thought that a belief that technical and scientific progress came from an assumption that man is essentially a tool making and using animal is wrong. He stated that 'tool making is but a fragment of biotechnics: man's total equipment for life. Basically he is more than a tool maker, he is a mind maker' [9].

But then when one looks at such design successes as computers or drugs, one suspects that with the increasing examining power of computers, much design will reduce to a Bunge model.

Such work bridges first, second and third order questions. Here one should look at Jacques Ellul and his 'Technological Society'. When technique becomes the central component of society, technology has replaced the former natural milieu and acquired the power to determine the ideas, beliefs and myths of modern man.

The ethical significance of building design is, I suspect, rarely argued for fear it will alienate potential clients. What values ought to direct building? Which values do direct building? Does engineering in building humanize because it gives more freedom to shape the building to any form? Or does it actually destroy humanity because it treats values as objects for manipulation?

It is argued that engineering, by satisfying all human wants, would do away with a lust for power. Romantics like Rousseau have argued that engineering is itself the outgrowth of a lust for power – which makes it evil. Or perhaps it is just neutral.

All of these questions are very important to us. We should try and debate them. Perhaps a single article written by a wise man every month in the Journal would be beneficial, answering one of the questions I have put, or ones other people can ask. It would not be Institutional policy but bearing witness to personal reflection. Such witness can help us all make choices about the work we do, our politics, our relationships, and, perhaps, withdrawing our cooperation from relationships which are unjust and exploitative. Thereby our work might be better understood.

I cannot end without commenting on two points which certainly would have interested some of our absent Gold Medallists. Hardy Cross and Nathan Newmark both came from the University of Illinois College of Engineering and would have been very interested in the current UFC research rating reviews. Cross was a great teacher and respected for it. He was charged with not publishing enough papers, although those he had were very significant even if not understood. In answer to threats, in a week Cross wrote a 10 page paper for the ASCE called 'Analysis of continuous frames by distributing fixed end moments'.

As Cross said 'there were no further discussions with the Dean of my being a Cross the College would not bear'. It might not have had the same effect today. It underlines what is obviously true about the current UFC system, which is a peer review to a fixed set of criteria determined, it appears, by scientists. Now peer recognition – within the university system – acclaim and reward is good. But peer examination can be very suspect, a conspiracy against the broader discipline. Lots of little academic clubs carrying out their day to day affairs within a framework of presuppositions about what constitutes a problem, a solution and a method. Kuhn claims such a background of shared assumptions makes up a paradigm and at any given time a particular scientific community will have a prevailing paradigm that shapes work in that field.

To have a paradigm shift – and engineering ought to be about paradigm shifts – look at Hardy Cross, for that matter look at Nathan Newmark's Latino-American Tower in Mexico – it involves bloodshed as in revolutions, admittedly only intellectual blood. Kuhn's argument is that the underlying issues, the judgement systems of the reviewers if you like, are not rational but emotional and settled not by logic and appeals to reason but by irrational factors like group affiliation and majority or 'mob' rule. That the UFC, and some universities, do not really encourage industry input into academic content and set up review boards of academics only is understandable, but not to be encouraged. It limits rather than expands, the scope of our discipline. Our Institution should take more of an interest in such review systems, it should demand the CIC does as well.

Felix Candela's hero is Ortega y Gasset and I must agree with him. Ortega contends that the perfection of the technology leads to a uniquely modern problem, not only are fewer technicians needed but there is a drying up of the imaginative or wishing faculty. In the past a designer was conscious of the things he was unable to do, of his limitations and restrictions. Years of energy were needed to realize a programme. Now with general methods to realize almost everything, man seems to have lost the ability to will any means at all. The traditional relationship between imagination and action has been reversed.

In the hands of a man devoid of imaginative faculty, engineering is 'an empty form like the most formalistic logic and is unable to determine the content of life'.

That fewer engineers are needed to carry out building structures should be dealt with by looking at those further problems in building engineering. The world needs more, not less building, and for it to be more energy effective, more affordable and so on. We must make the most of our closer relationships with CIBSE – so that we can tackle similar problems together.

That imagination has declined should be resisted by our taking an interest in the philosophy of engineering. Ortega's view was that we should look to Eastern construction. Probably it lies in our accepting – and being seen to accept – our share of responsibility for the world's construction needs. Then perhaps we can take the opportunities Paul Valery talks about:

'Our fine arts were developed, their types and uses were established, in times very different from the present, by men whose power of action upon things was insignificant in comparison with ours. But the amazing growth of our techniques, the adaptability and precision they have attained, the ideas and habits they are creating, make it a certainty that profound changes are impending in the ancient craft of the Beautiful. In all the arts there is a physical component which can no longer be considered or treated as it used to be, which cannot remain unaffected by our modern knowledge and power. For the last twenty years neither matter nor space nor time has been what it was from time immemorial. We must expect great innovations to transform the entire technique of the arts, thereby affecting artistic invention itself and perhaps even bring about an amazing change in our very notion of art.'

Paul Valery, Pièces Sur L'art,
'La Conquête de l'ubiquité', Paris

References

[1] *The Shorter Oxford English Dictionary* (1978) Oxford University Press, Oxford.

[2] Happold, E. (1992) The Nature of Structural Engineering. *The Structural Engineer,* **70** (20).

[3] Engineering, the heroic art: Arnold Pacey, in *Great Engineers and Pioneers in Technology,* Vol 1, (eds. R. Turner and S.L. Goulden).

[4] Mitcham, C. and Mackay, R. (eds) (1983) *Philosophy and Technology,* The Free Press, Macmillan Publishing Co. Inc.

[5] Rasmussen, S.E. (1959) *Experiencing Architecture,* The M.I.T. Press, Cambridge, Mass.

[6] Kuhn, T.S. (1962, 1970) *The Structure of Scientific Revolutions,* University of Chicago Press.

[7] Galbraith, A. (1992) *The Culture of Contentment,* Houghton Mifflin Company.

[8] Bunge, M. (1967) Scientific Research 11: The search for truth, Vol 3, Part 2 of *Studies in the Foundations, Methodology and Philosophy of Science,* Springer Verlag, Berlin.

[9] Ochser, P.H. (ed) (1966) Technics and the nature of man: Lewis Mumford in *Knowledge Among Man,* Simon and Schuster, New York.

4
Some highlights in the history of bridge design

Anthony Flint

Anthony Flint's early training was with the Great Western Railway. After graduating from the Imperial College, London, he engaged in aircraft structural research at the Royal Aircraft Establishment. Following a Postgraduate degree at Bristol University, he remained as a university Research Fellow until 1953, working on ship structures. He was a Lecturer and subsequently Reader in structural steelwork, at the University of London from 1953 to 1972.

He was a member of the Committees of Inquiry into the Clyde Crossing and Emley Moor mast collapses. He served on the Government Committee of Inquiry into the failure of Milford Haven and Yarra bridges, 1970–1973, and was joint author of the *Interim design rules for box girder bridges*.

He is a member of several British Standard and International Code committees on loads, bridges and steel structures.

In the firm which he founded in 1958, he has directed the design and supervision, appraisals and special investigations for bridgeworks, towers and masts, offshore structures and buildings. He has specialized in structural aerodynamics, reliability analysis and stability problems.

Introduction Bridge design is a mental process of synthesis, aimed to achieve a solution to bridging a gap. In pre-history it consisted only of adapting natural phenomena by using stones or logs as cantilevers or beams, or by stringing vines between cliff faces or trees.

In the course of early civilization the process began to develop by empiricism, making use of experience and trial and error. (If early bridges did not fall down during construction they were usually able to support the relatively insignificant loads from man and beast.)

The evolution of the arch as the structural form for all the major bridges until 200 years ago was accelerated by the

Structural Engineering: History and development, Edited by R.J.W. Milne. Published in 1997 by E & FN Spon, London. ISBN 0 419 20170 X.

discovery of the laws of geometry by the great Greek mathematicians, such as Euclid, which led to the adoption of circular, ellipsoidal or parabolic forms which have persisted to this day. Such forms have, either by good fortune or by familiarity, proved pleasing to the eye. With the later creation of new structural materials designers have increasing need to consider aesthetics.

As man-made materials were developed and ambitions to span larger chasms grew, the synthesis has rapidly expanded to encompass the use of knowledge of properties of materials, of mathematics of algebra and the calculus, of static and dynamic loading (both man-made and natural), of new constructional techniques and of the economics of construction. It has to overtly consider performance requirements, including those of safety and durability, which are increasingly onerous.

In the past 60 years design has passed from the hands of a relatively few gifted all-rounders to multi-disciplinary teams providing the wide variety of expertise now needed to tackle the many facets of the now complex process. The need to transfer knowledge in increasing volume, gained by experience (in triumph or disaster), original thought or research has led to ever changing codified rules with which the teams must keep abreast. This is similar to the reliance in earlier times on rules based on well-tried practice before the theory of structures was available. In the intermediate periods engineers relied on personal contacts and reading of the relevant important papers to practise the art, which is still essential for progress.

Structural forms

The structural forms suitable for bridging have changed little since early times. The stone clapper bridges and the modern concrete viaducts are of the same family of beam structures. The stone arches, evolved from Sumerian times at least 6000 years ago, and so perfectly executed by the Romans in the Pont du Gard in 19 BC (Fig. 4.1) were the progenitors of the Sydney Harbour Bridge (Fig. 4.2). The Forth Rail Bridge (Fig. 4.3) is a descendant of the primitive cantilevered structures in stone and timber. The suspension bridges of natural ropes or the earliest iron chain bridge over 2000 years ago, attributed to the Chinese, have evolved in our times to the Tsing Ma Bridge, recently constructed (Fig 4,4).

The truss form is reported to have been used in wooden trestle bridges as early as 55 BC and by Apollodorus in a crossing of the Danube used by Trajan in 104 AD. Queen- and King-post types were developed in the sixteenth century by

Figure 4.1 Pont du Gard, 19 BC. (Source: The Institution of Civil Engineers.)

Palladio, but it was more than two centuries later before there was a spawning of the Howe, Pratt, Warren, Whipple, Fink, Bollman and other systems, which resulted from the production of man-made tensile materials and development of methods of connection. Many of these were of redundant form and were introduced before accredited methods for their analysis were available. They were used as both under- and through-girders, many in the USA being covered for comfort and durability. These unnatural forms, although often economical, are commonly considered to be unattractive.

The unbraced vierendeel truss, developed in Belgium in the early twentieth century, has not proved to be popular for other than footbridges and is usually uneconomical. However, the crossframes of the Kap Shui Mun Bridge, designed by Leonhardt and Andra, are of that form to facilitate uninterrupted passage of traffic in the lower deck.

4.

4

Figure 4.4 Tsing Ma Bridge, designed by Mott MacDonald, 1997.
(Source: Kvaerner Construction Group Ltd.)

Developments have frequently relied on a combination of forms. The first suspension bridges using wrought iron chains, such as the Winch Bridge over the Tees, built in 1741, had flexible flooring. Even Telford's Menai Straits Bridge, built in 1825, was unstiffened. Judge Finley's patented chain suspension bridge designs in the early 1800s incorporated a timber beam-type deck to distribute local loads, of which the Jacob's Creek Bridge was an early example (Fig. 4.5). This combination was supplemented by cable staying, both from the towers and from underneath which also provided stability in wind. The need for such stiffening grew with increase in traffic loads, and was fully appreciated by Roebling in his successful and radical design for the Niagara Railway Bridge, completed in 1855 (Fig. 4.6). The relative stiffness of the decks gradually increased to such an extent that in Stephenson's Britannia Bridge (Fig. 4.7), the forerunner of modern box and plate girder construction, the suspension system of its original design was omitted as unnecessary following Fairbairn's experimental work. With increasing spans and total weight, and

Figure 4.2 Sydney Harbour Bridge, designed by Freeman, 1932.
(Source: The Institution of Civil Engineers. Photo: W. Ingle.)

Figure 4.3 Forth Rail Bridge, designed by Baker, 1889. (Source: The Institution of Civil Engineers.)

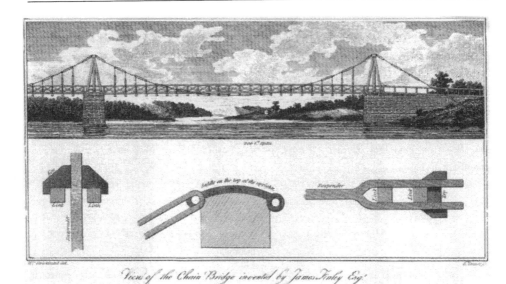

Figure 4.5 Jacob's Creek Bridge, designed by Finley, 1801. (Source: The Institution of Civil Engineers.)

consequent gravity stiffness, the need for stiff decks for suspension bridges has diminished. This was understood by Ammann in his design of the George Washington Bridge of 1067m span, completed in 1931, which, until the addition of a second deck, had no stiffening girder. There have been recent examples of 'ribbon' bridges in which the sagging deck is the suspension element, reverting to the earliest primitive form.

Similarly resulting from the advent of new materials with tensile strength are the cablestayed bridges which, since 1945, have proved increasingly efficient for longer spans, the Pont de Normandie being the current front runner with a span of 856m. These were really developed from the earlier suspension systems, but eliminate the catenary cables which are not very useful for local loading, but retain the straight stays which were a feature of many of the earlier suspension bridges, such as Ordish's Albert Bridge over the Thames, which used the catenary chain primarily to prevent sagging of the inclined stays. Their penalty is the compressive load induced in the deck which will ultimately govern their limits of application.

The arch form was also combined with others – Palladio built a trussed timber version. The tied arch was first used by

Figure 4.6 Niagara Bridge, designed by Roebling, 1855. (Source: The Institution of Civil Engineers.)

Palmer, Wernwag and Burr in the early 1800s. The supreme final work of Brunel at Saltash, opened in 1859, followed his earlier development for the Usk Viaduct and Chepstow Bridge and incorporated a masterful combination of tubular arch, truss, beam and suspension using the chains intended for the Clifton Bridge (Fig. 4.8).

Forms of bridge support have also changed little since early times, although the techniques for providing it have advanced radically. Piling and caissons were used by the Romans, but the development of motive power and the use of compressed air have vastly increased their capabilities. The use of pontoon supports by the Assyrians in the ninth century BC and by Xerxes to move his army over the Hellespont in 480 BC was

Figure 4.7 Britannia Bridge, designed by Stephenson, 1850. (Source: The Institution of Civil Engineers.)

Figure 4.8 Saltash Bridge, designed by Brunel, 1859. (Source: The Institution of Civil Engineers.)

Figure 4.9 Tower Bridge, designed by Wolfe-Barry, 1894. (Source: The Institution of Civil Engineers.)

followed by the Mulberry Harbour in 1944, using that wonderful kit for truss bridges designed by Bailey.

Developments in mechanical engineering have led to the feasibility of mobile supports facilitating navigation spans without the need for massive approach works and gradients. The swing, lifting and bascule bridge evolved, of which Tower Bridge is probably the most famous (Fig. 4.9). The modern military tank-launched assault bridges exemplify a combination of mechanical and structural systems. The transporter bridges, such as that at Newport, are ingenious combinations.

Materials and structural components

Design is always related to the materials available for bridge construction, the development of which has often been stimulated by designers' demands. The Romans made the best use of their timber, using alder for piles and fir, cypress and cedar for their superstructures. Most of their early arch bridges used Tufa or Travertine limestone, the former requiring stucco protection for durability and the latter having the greater compressive strength. They also advanced the manufacture of bricks from clay, earth and sand for use in bridgeworks. Their discovery of the process of making pozzolana cement from volcanic clay burnt with lime provided the means to make concrete which could set underwater. Although they did not use these materials for most superstructures, they built some concrete arch bridges at the end of the Empire.

Vitruvius's descriptions of Roman practices were rediscovered in the late fifteenth century. However, it was not until Smeaton adopted the use of pozzolana cement for use in the foundations of the Eddystone Lighthouse that the material became used extensively in bridgeworks. By 1811 Aspdin had established his first plant to make artificial cement by burning clay with chalk, patented as 'Portland cement' in 1824. The first concrete bridge, following Monier's patents, was built in 1875. Since then the products have developed with high-alumina cement for rapid hardening, high-silica, essential to reduce problems of high heat of hydration which accompany large pours, and chloride-resistant cement for durability.

Reinforcement of concrete was proposed by Dodd in 1808 and was later used by Telford in the Menai Bridge towers. Notable experiments were undertaken by Hyatt in the 1850s and the first concrete beam theory was propounded by Koenen in 1886, followed by Coignet's elastic theory which is still in use.

By 1892 Hennebique had developed the T-beam principle and, by 1900, had designed at least 100 concrete bridges, many being of arch form. These were followed by developments by Considere and Melan, the latter devising self-supporting systems of reinforcement. The great achievements of Maillart started in 1901 with his bridge at Engadin which was of box form with an arch span/rise ratio of 10. By 1933 he had designed the Schwanbach Bridge which is considered to incorporate the best in structural analysis and aesthetics.

The use of reinforcement of higher strength created problems of cracking which were overcome by the introduction of prestressing by Freyssinet, whose Esbly Bridge in 1949 was significant. His designs have been followed by many notable prestressed bridges, that by Maunsell for the Gladesville Bridge in 1964 being the first concrete arch of span in excess of 1000ft.

The advent of iron smelting with coke, begun by Abraham Darby in 1713, was a momentous advance in production of materials for bridge building. Cast iron produced in this way was first used for a bridge by Darby's grandson and Wilkinson in the construction of the Coalbrookdale Ironbridge which, being an arch, made best use of its compressive strength (Fig. 4.10). (It was designed by an architect, Thomas Pritchard!)

Figure 4.10 Coalbrookdale Bridge, designed by Darby, 1779. (Source: The Institution of Civil Engineers.)

Wrought iron was first made in commercial quantities by Henry Cort in 1784, although it was little used until 1850. Rolling mills developed in the early 1800s, and these eventually provided the designer with a wider range of products for bridge use. Plate girders were being made by the 1830s, as were rolled angle sections. I-sections were first rolled in 1844. Of equal importance was the production of rivets which made connections possible.

Many bridges in the early nineteenth century combined the use of cast iron in compression members with wrought iron in tension parts, as in the Whipple and Howe Bridges. The same principle has continued in the use of steel/concrete composite construction today.

The first cable suspension bridges were erected in America and Scotland in 1816 and the first wire cable bridge for public use was built in Geneva by Dufour in 1823. By 1855 wrought iron wire was available to Roebling for the Niagara Bridge in which the wires were compacted into a circular cable and wrapped with soft wire for protection. By the time Roebling's Brooklyn Bridge was ready to build in 1883, galvanized wire was available.

The invention by Bessemer in 1855 of the production of wrought steel by the blast process and the subsequent building of the converters opened the way for modern bridge design. Steel began to be used in the 1870s, for example in Eads' St. Louis Bridge in 1874. The later processes of manufacturing steel tubes and the advent of welding provided designers with a choice of new efficient sections which have been used in a number of bridges. Since that time the properties of structural steel have been improved by alloying or heat treatment to provide higher yield strength with ductility and weldability needed to reduce bridge weights and achieve greater spans. The science of arc welding followed, but was not in common use in bridgework until the 1930s.

The use of structural aluminium alloys for bridges, thought in the 1950s to offer designers low weight/high strength materials with great potential, has, for reasons of cost of materials and fabrication, not yet been widespread. It has however been successfully adopted in British military bridging.

Bridges of composite plastics have been designed by Maunsell and the first of these, a footbridge with glass-fibre reinforced deck and Kevlar stays, has recently been built at Aberfeldy (Fig. 4.11).

Analysis of structural systems and their parts

The concept of structural statics was propounded by Greek philosophers, such as Aristotle in the fourth century BC. However the foundation of mechanical science is attributed to Leonardo da Vinci, who was educated in the Greek work, in the fifteenth century. It was Galileo who, in 1638, first expounded theories of structural analysis – including those for stresses in frames and strains in beams. These were later corrected by Hooke, who, in the late seventeenth century, discovered the basic law of elasticity on which so much of design is still based. He understood 'the plane-sections remain plane' hypothesis propounded by Navier in 1826, and developed by his contemporary, Mariotte. Navier is credited with the establishment of the theory of elasticity following his Leçons in 1826. Hooke also outlined the principles of the catenarian arch, which was developed by Gregory, Bernoulli, Coulomb and Young over the following century.

The art of building arch bridges was revived by the Fratres Pontifices in the eleventh century, and the Pont d'Avignon, due to St. Bénezèt, followed. It was la Hire who, in 1695, first scientifically applied statics to analysis of arches. He used the funicular polygon on which Perronet later based his design

Figure 4.11 Footbridge at Aberfeldy, designed by Maunsell, 1993. (Source: G. Maunsell & Partners.)

formulae. Potential arch failure modes were demonstrated by tests by Danisky. These were used by Coulomb in developing his theory of 1773 in which he recognized the importance of friction in providing stability. That work was extended by Lame, Clapeyron and Navier. By 1791, Perronet, in his last great design for the Pont de la Concorde had, by theory and testing of materials, achieved a low rise arch form and the understanding that intermediate piers could be more slender than those at the ends, similar to his earlier Pont de Neuilly (Fig. 4.12). By 1820 Young had advanced arch theory to a stage from which it has only required refinement.

L'Ecole des Ponts et Chaussées, first directed by Perronet, accompanied by the work which stemmed from l'Ecole Polytechnique by Monge, Fourier, Lagrange, Poisson and others, led the way in the design of bridges in the early nineteenth century. Navier, who had been a pupil of Perronet, with the aid of Moseley in England, developed the theory of continuous beams which was put into effect in the design of the Britannia Bridge by Stephenson – the first example of prestress by con-

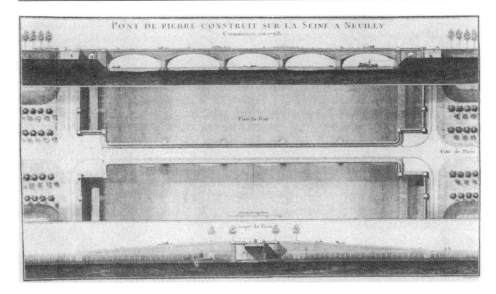

Figure 4.12 Pont de Neuilly, designed by Perronet, 1768. (Source: The Institution of Civil Engineers.)

trolled initial lack-of-fit of supports. The theory was developed by Clapeyron while he was in St. Petersburg and was extended by Winkler in 1862, who, with interest in railways, also analysed long beams on elastic foundations, later developed by Hetenyi and others.

The discovery of the mathematics of the infinitesimal calculus by Leibnitz and Newton at the end of the seventeenth century opened the door to much of later structural theory. This was applied by the Bernoulli brothers to calculate the deflections of beams and put to good effect by Euler in 1757 in the analysis of the buckling loads of perfect struts, following his tuition at the Leibnitz mathematic school. His work was developed by Young for imperfect and eccentrically loaded columns, from which stemmed theories of elastic stability on which designs of many modern bridges depend.

The efficient use of iron and steel has much depended on the avoidance of buckling in some form. In the nineteenth century reliance was placed on empirical rules and testing of both scale models and full scale testing of components, such as that undertaken by Fairbairn, with the help of Hodgkinson, for the Conway and Britannia bridges which resulted in the adoption of compression chords of cellular form. It was not until the early twentieth century that the theories of elastic stability began to emerge. By 1946 the Cologne-Deutz steel box girder bridge had been designed, followed by the plate girder Bonn-Beuel Bridge two years later.

Stability increasingly became a problem for designers, as witnessed by the collapse during erection of Cooper's Quebec Bridge in 1907, due to buckling of compression chords, and the later box girder collapses in our times. Practical rules for strut design, based on tests such as those due to Tredgold and Hodgkinson, and later improved by Rankine on the basis of Gordon's formula, were all based on experimentation, although the first reliable tests were undertaken by Bauschinger and Tetmajer in about 1887.

There were a number of failures of half-through iron bridges due to compression chord buckling which were investigated by Jasinsky, who developed the 'strut on elastic foundation' analogy which is still the basis for the rules in BS 5400.

The twentieth century has seen extensive theoretical and experimental work on buckling, which was initially based on the differential equations of stability, and later used approximate energy methods due to Rayleigh's work on vibrations. Non-linear finite element computerized analysis has now vastly extended our capability to tackle stability problems. The contributors to the advancement of knowledge in this field are numerous, including those engaged in the development of aircraft design from the 1920s onwards. Engesser, Wagner, Timoshenko and Bleich deserve special mention, bridge designers owing much to their work. The research which followed the collapses of parts of the Milford Haven, Yarra and Koblenz box girder bridges, and Horne's work in developing new design rules radically improved our knowledge of behaviour of steel plated structures.

Early truss design was by rule of thumb without analysis, and commonly all the bracings were of the same size in a structure. The triangle of forces was discovered by Stevin in 1586 by experiments on loaded strings. Navier is attributed with the first precise analysis of statically indeterminate systems. However, it was Squire Whipple who first analysed the stresses in his trusses in 1847, and Bow published a treatise on truss bracing in 1850. The methods were established in Rankine's classic text in 1858, *Applied Mechanics*, which has served bridge designers to this day. There were considerable

developments in the mathematical analysis of statically indeterminate systems by Rankine, Maxwell, Ritter and others in
the 1860s and 1870s, and in parallel, developments in graphi-
cal static analysis methods by Rankine, Culmann, Mohr,
Fidler, Müller Breslau and Williot which were to prove popular with bridge designers. The matter of 'secondary' stresses
due to rigidity of connections was addressed by Manderla who
provided the equations of equilibrium at joints. Approximate
methods for treating these were later devised by Mohr.

The concepts of the principle of the balance of energy were
used by Euler and proved by Castigliano in his principle of
least work, which he used to calculate the deflections of
beams. These concepts have provided researchers and designers with some of the most powerful and simple weapons in the
solution of structural problems which are used today.

By the time of Brunel he was able to calculate the catenary,
catenary of uniform strength, (due to Gilbert), and the parabolic form for the Clifton Bridge. Rankine in 1858 published
the theory for inclined- as well as that for two- and three-
hinged stiffening systems. It was not before 1888 that the
elastic and deflection theories were propounded by Melan
which were used in many of the designs of the great bridges
that followed. They were first used by Mossieff in the calculations for the Manhattan Bridge, and later for the George
Washington, and Golden Gate bridges. The basic differential
equations of the deflection theory were solved by Timoshenko
in 1928 by use of Fourier series. In 1939 Southwell extended
his relaxation theory to suspension bridges which was applied
in practical form by Crossthwaite in the design of British
bridges. The process has led, with the invention of the micro-
chip, to the current computerized methods of analysis, based
on the tenets of equilibrium and displacement compatibility,
expounded by Brotton in 1963 and others. These now provide
the designer with tools with which to analyse the load effects
in all parts of a bridge due to loading in three dimensions,
even, by use of fusible links, allowing for hangers slackening.

The deflection theory was linearized for simplicity by
Godard, Bleich, Hardesty and Wessman and Pugsley (who
propounded an influence coefficient method and a 'beam on
elastic foundation' analogy). These have provided simplified
methods of preliminary design analysis.

**Strength of
materials**

The mechanical properties of materials have been investigated
by testing since the seventeenth century, first for timber by

Mariotte in the 1670s. Reaumur used mechanical testing in 1722 to study the tensile properties of steel and Musschenbroeck provided data on the properties of both wood and steel in 1729 and confirmed Euler's theory for struts. In 1733 Coulomb tested the tensile and shear properties of sandstone and Perronet supported his designs by tests on stone.

The industrial revolution provided the powerful machines for material and component testing which provided designers with data on the properties of materials, and the strength of components and their connections, well used by Barlow, Telford, Fairbairn and others.

The advent of mild steel was followed by unexpected failures in riveted structures by brittle fracture by 1886. Such behaviour of steels at certain temperatures was highlighted by the failure of three vierendeel trusses in Belgium by 1940, of the Liberty ships in the war and by the collapse of the King's Street plate girder bridge in Melbourne in 1962.

Fatigue became a problem in iron railway axles and there were failures in a number of early American iron trusses due to it. Fairbairn undertook repeated load tests on tubular beams which led to the appreciation of the significance of stress range and the concept of the endurance limit. Scientific study of the phenomenon was begun by Wöhler in 1847, primarily for application to rolling stock design. By 1874 Gerber had developed a relationship between fatigue life and stress range.

There have been extensive investigations into the physics of these phenomena since 1948 and the science of fracture mechanics has followed, which provides designers and fabricators with new rules for avoidance of the risks.

The problems with aluminium alloys of stress corrosion and the loss of strength due to welding have been the subject of much recent study.

Loading and load effects

Until the advent of railways the live load had been a small part of the load to be carried by bridges, and it then became important to consider not only the static effects but also dynamic amplification due to the transit of traffic, and later, the hammer blow effects from locomotives with which the classic work of Inglis was concerned. In recent times highway traffic loadings have escalated, and statistical theory has been used to predict extreme loadings for design.

With increase in spans, loading due to wind became of importance, and, following the ill-famed Tay Bridge collapse in 1879, attention was focused on the pressure due to wind that

Figure 4.13 Tacoma Narrows Bridge, 1940. (Source: The Institution of Civil Engineers.)

should be allowed for in design. Tests by Fowler and Baker on wind pressure for the Forth Railway Bridge led to the unnecessarily high design value of 56psf. Since that time the methods of wind load prediction have developed on the basis of long periods of recording of wind speeds, using statistics, the measurement of aerodynamic coefficients in wind tunnel tests and the radical improvements in treating the dynamic response to natural wind, largely due to Davenport. These have been accompanied by remarkably accurate methods of calculating dynamic characteristics of bridges.

Problems of aerodynamic instability of suspension bridges were encountered by the Victorian engineers, which was the cause of the Brighton chain pier failure in 1836, and that of the Wheeling Bridge in 1854. They were combatted by staying and deck stiffening, and diminished as bridge weights grew. This caused designers to forget past disasters until the collapse of the Tacoma Narrows Bridge in 1940 (Fig. 4.13). Not only was the deck width of that bridge small, but its stiffening girder was of plate girder construction (in contrast to the earlier truss stiffening) and had a depth/span ratio of 1/350 compared with 1/40 of earlier bridges.

Since that time experimental and theoretical studies, with the practice of wind tunnel testing of models of all major

Figure 4.14 Severn Bridge, designed by Roberts, 1966.

bridges, as used by Fraser and Scruton for the Forth and Severn bridges designed by Roberts in the 1950s, have provided designers with the basis for avoiding unacceptable vibrations due to vortex excitation and flutter. The aerofoil Severn Bridge box girder deck was revolutionary in design concept, providing low drag and, by its high torsional stiffness, stability (Fig. 4.14).

Design of bridges in earthquake zones has also benefitted from the developments in analysis of dynamic response to spectra of ground movements and in prediction of magnitudes of acceleration. The recent collapses of sections of elevated highways in California show that there remains scope for improvement in design practices, particularly in detailing.

With increasing spans there is more incentive to build bridge supports in navigable waters, and this has brought the hazard of failure due to ship collision. While knowledge of forces due to ship impact remains imprecise, many designs now have to account for this risk, either by providing protection or by designing supports to resist possible impact.

Foundations and construction techniques

Such an abbreviated history as this cannot adequately cover the extremely important influences on design, which are not apparent to the layman, of foundation design and methods of construction and erection.

Although the basic foundation types are simple, the advent of mechanical power for pile driving and the use of compressed air in caisson construction have contributed towards our capability to bridge large spans with a wide variety of ground conditions at supports. Moreover, the developments in the understanding of soil and rock mechanics in recent years have provided designers with scientific bases for deriving safe loads for foundations and anchorages. The design against scour from rivers, which has caused so many failures, has evolved with our knowledge of hydraulics.

What is possible to design for economy depends on the current techniques available, and in earlier times these relied on the carpenters to provide centreing for arches. The feasibility of designing many of our modern bridges depends on not only the mechanical power, which enabled the Victorian engineers to lift their bridges into position, but also methods of cantilevering, with the aid of prestress for concrete bridges, incremental launching, cable spinning and slip forming of concrete towers.

Bibliography

For those interested in the detailed history of the art, the literature below will provide references to many of the important contributions.

Birdsall, B. (1988) Footnotes to suspension bridge history prior to 1950. Paper submitted to the 1st Oleg Kerensky Memorial Conference.

Brown, C.W. (1988) Evolution of analysis for cable supported bridges. Paper submitted to the 1st Oleg Kerensky Memorial Conference.

Brown, D.J. (1993) *Bridges: Three thousand years of defying nature*, Mitchell Beazly.

Charlton, T.M. (1982) *A History of the Theory of Structures in the 19th Century*, Cambridge University Press.

Chatterjee, S. (1991) *The Design of Modern Steel Bridges*, BSP Professional books.

Gifford, E.W.H. (1962) The development of long span prestressed concrete bridges. *Structural Engineer*, **40**(10) October.

Godfrey, G. Bernard (1957) Post war developments in German steel bridges and structures. *Structural Engineer*, **35**.

Hopkins, H.J. (1957) *A Span of Bridges*, David & Charles.

Johnson, J.B., Turneare, F.E. and Bryan, C.W. (1893) *The Theory and Practice of Modern Framed Structures*, 1st edition, Wiley.

Kemp, E.L. (1979) Links in a chain; the development of suspension bridges. 1801–70. *Structural Engineer*, **57A**(8) August.

Kerensky, O.A. (1959) Long-span suspension bridges. Maitland lecture. *Structural Engineer*, **37**(7) July.

Leonhardt, F. (1982) *Bridge Aesthetics and Design*. Deutches Verlags-Anstatt.

Pannell, J.P.M. (1964) *An Illustrated History of Civil Engineering*, Thames & Hudson.

Parsons, M.F. (1988) Major suspension bridges completed since 1960. Paper submitted to the 1st Oleg Kerensky Memorial Conference.

Pugsley, A.G. (1957) *The Theory of Suspension Bridges*, Edward Arnold.

Rankine, W.J.M. (1851) Mechanics (Applied), Encyclopaedia Brittanica.

Rolt, L.T.C. (1957) *Isambard Kingdom Brunel*, Longmans Green.

Steinman, D.B. and Watson, S.R. (1941) *Bridges and their Builders*, G.P. Putnam's Sons.

Straub, H. (1952) *A History of Civil Engineering*, Leonard Hill.

Sutherland, R.J.M. (1988) Some aspects of the history of suspension bridges. Paper submitted to the 1st Oleg Kerensky Memorial Conference.

Timoshenko, S. (1953) *History of the strength of materials*, McGraw Hill.

5
Working on the edge – the engineer's dilemma

Sir Jack Zunz

Ove Arup & Partners

Jack Zunz is a consultant to the Ove Arup Partnership. He joined Arups in 1950 and became successively a Senior Partner, Chairman of Ove Arup & Partners and Co-Chairman of the Ove Arup Partnership in the UK and the Arup group of practices worldwide. He has been involved with many well-known projects in the UK and elsewhere. He was awarded the Institution Gold Medal in 1988 and knighted in 1989. He is the author and co-author of a number of publications.

Engineering is essentially a social art – the science part is taken for granted and it gets more than enough exposure. Being a social art practised by professionals, there is an automatic implication of responsibilities and duties, particularly to clients and to society at large. That such responsibilities exist is beyond argument, what is more difficult to establish is their extent and their limits.

This Institution, like other professional societies and institutions, has a code of conduct for its members, a code which is often adhered to more in its breach than in its observance. Yet there *is* a code and its importance should not be underestimated, so that today when we are celebrating the 60th anniversary of our Charter it is appropriate to examine this whole issue of ethics and morals in so far as they are relevant to our activities as structural engineers.

One of the most distinguished presidents and a Gold Medallist of this Institution, Oleg Kerensky, chose to speak about the engineer's ethics in his presidential address, an address which on re-reading more than 20 years later is as thought-provoking and as relevant today as it was two decades ago [1]. Kerensky spoke about the engineer's place in the spectrum of 'so-called' professional people. I say 'so-called' because he asserted that chartered engineers' claims to belonging to a profession did

Structural Engineering: History and development, Edited by R.J.W. Milne. Published in 1997 by E & FN Spon, London. ISBN 0 419 20170 X.

not satisfy the criteria proposed by Bennion in an interesting book called *Professional Ethics* [2]. Bennion's standards for those deserving professional status were rigorous. Kerensky was clearly equivocal about the engineer's right to call him or herself a 'professional person'.

Since delivering his presidential address the semantics of the word 'profession' have become looser. But even more fundamental has been a systematic whittling away of what was perceived to be the professions' privileged and protected position in our social and economic life. So whether or not Kerensky was correct in stating that our 'so-called' profession did not meet the Bennion criteria, we, along with other perhaps more accepted professions, have been relegated from the Premier League.

In today's context it is little comfort to recall that in 1968 Lord (R.A.) Butler said:

'Society in the future may become progressively intolerant of voluntary professional institutions especially if they are the bulwark of private practice, and yet be oblivious to the truth that in these institutions resides a most precious liberty essential to the health of a civilised society' [3].

It is unlikely that Butler was sufficiently prescient to equate his vision of 'society' with the attitudes of his own political successors, who have launched a systematic assault on the professions under cover of accusations of closed shops, with all the restrictive practices and monopolies that these imply. So we have moved on from Kerensky's time and we have to accept that for the present anyway, we are merely providers of services to be bought and sold in the marketplace like other goods and, equally unfortunately, other services akin to ours. The pendulum will one day swing away from this situation and back to common sense, but it will take time. In the meantime life must go on and we must do the best we can under the given circumstances.

Recognition by others of what we consider to be our proper place in society is interesting, but what is much more important and to the point is how we conduct ourselves and how we are seen by others as living up to our own aspirations.

There are, then, other aspects of professional ethics and morals which Kerensky did not touch on. Certainly we must all strive to be clean, honest and polite – clean, honest and polite not just in the narrow sense but in the broadest possible interpretation of the meaning of these words. Care about the environment, fair and honourable dealings in our personal and professional lives and concern for our neighbours and humanity in general are all essential constituents of professional ethics. But more difficult to articulate are our duties to our clients in particular and to society in general, and these can sometimes be in conflict. It probably requires little deliberation to conclude that to work on the design or construction of concentration camps would be unacceptable. Some colleagues refuse to work in this or that country because of its politics, or more seriously, its oppressive regime and disregard for human rights. Others again object strenuously and conscientiously to anything to do with war, the military and associated works. This is all perfectly reasonable and understandable and causes no more than practical dilemmas. Life becomes more difficult with other pressure groups. For instance, animal rights campaigners would not wish to be associated with laboratories where experiments on animals might be carried out. One can go on and imagine a whole number of situations which might offend the personal principles and commitments of the engineer, and where he or she may not wish to be involved. But, rational or not, there is usually a plausible explanation and reason for not wishing to practise one's professional skills.

What is more difficult to establish (if indeed it is at all possible) is when to say 'NO' to a request to design a structure which, in the opinion of the designer, is inappropriate, is unnecessarily expensive, is too difficult to build, in other words is an apparently wilful demand of the client. There is also the dilemma that the designer may wish to achieve certain self-indulgent, self-promoting or other ends in fulfilling the client's brief and in so doing move uncomfortably close to the edge. This is as relevant today, when we work as structural engineers with architects who more often than not are the prime designers responsible to a promoter, as it was formerly when the mason, carpenter, engineer, or architect – call them what you will – was the master builder/designer.

Architecture has several faces. The functional and the economic are generally understood and are subject to rational examination and debate. It is the artistic face which causes problems. Architecture *is* an art – the mother of all arts. Goethe referred to architecture as 'frozen music'. The artistic quality of architecture is fundamental to construction, indeed it is the essence of much of our culture and civilization. But – and there is a 'but' – what are those special features which go to make up the artistic quality of architecture and more particularly what is the role of technology in all this? There is no simple answer – if indeed there is an answer at all.

Whether or not the engineer is independent, whether or not he is the prime designer, or working with an architect, or indeed whether he is the sole designer, how can one judge what is sensible or reasonable – in short how far should one go?

There are apparently no rules nor is there a deterministic answer. But it may be instructive to examine some historic as well as some contemporary references.

In discussing moral and ethical issues the question of structural honesty comes into contention. There have been and there still are many proponents of the idea that for every architectural/structural problem there is a 'correct', a 'right' and an 'honest' solution. Structural honesty has been linked with the functionalism of the modern movement in architecture with all its social and moral overtones.

There is no such thing as structural honesty. There are no 'correct' solutions, no 'right' answers, but just to make certain that there are no misunderstandings I am not referring to the analytical or mathematical content of our work, though here too there is often scope for debate. I am referring to the *choice* of structure, its shape and content. There are objective reasons why some solutions, some structures appear better than others; they may be more economic, they may be thought to last better, and they may fulfil their function better than others. However, even the judgements on function enter the subjective territory which soon merges into fashion, into state-of-the-art expertise, conventional wisdom and personal prejudice or preference.

It is worth touching briefly on some historical landmarks. The designer of buildings has historically (except for the last century or so) generally been self-reliant with respect to structural expertise. Although structural engineering as a specialization, as we know it today, is a comparatively recent development, the issues facing the designer on how far to go, how far to stretch technology, the client's purse, the support of the public, are not new.

George IV as Prince of Wales, six generations before the current Prince, leased a small house in what is now Brighton, in the Steine, a valley running down to the sea. He commissioned Henry Holland, an architect, to enlarge his house into a pleasant classical building with a domed rotunda. However, it was after his father George III's mind failed in 1812 and the

Figure 5.1 Brighton Pavillion exterior. (Reproduced by permission of The Royal Pavillion, Brighton. Open daily.)

Prince of Wales became Prince Regent that the great reconstruction of his house began. He commissioned Nash, who transformed *inter alia* the exterior (Fig. 5.1) by adding to Holland's house the onion-shaped domes, tent-like pavilion roofs and pinnacles and minarets in the style of a mogul's palace – and all this in Brighton! The cost was not insubstantial.

This royal extravaganza, though sold by the monarchy to the town of Brighton in 1845, has over the years become a much loved, if eccentric and odd, attraction to millions of visitors. What is of particular interest to us and not apparently obvious is that Nash's extrovert response to his royal patron is no mere decoration, but is supported and driven by appropriate structural forms (Fig. 5.2).

Nash claimed that he did not depend on designs produced by the iron founders of the day, though his claims should be treated with some circumspection. He was getting on in years and was about to be dismissed from his position as architect for Buckingham Palace. The Prime Minister, The Duke of Wellington, wanted to make a 'hash of Nash' [4] for having sanctioned unauthorized expenditure as well as for designing potentially unsafe floor beams. Nash may also not have given due recognition to the work of J.U. Rastrick, one of the leading iron founders of the day. But whether or not Nash was personally responsible for the structures of his buildings and whether Rastrick or others were party to the design, the structure itself

Figure 5.2 Section through Brighton Pavillion. (Reproduced by permission of The Royal Pavillion, Brighton. Open daily.)

is clearly an integral, if not an essential part of this most popular and successful folly.

Moving back in history to the fifteenth century it is worth looking at what must be one of the most significant landmarks in structural engineering, Brunelleschi's dome in the cathedral of Santa Maria del Fiore in Florence (Fig. 5.3). The design and construction of the dome has been described as the work which gave birth to scientific engineering in the Renaissance [5, 6]. Placed in the context of early fifteenth century construction expertise, it must be regarded as one of the greatest building achievements of all time. One can only speculate on the passion, the ambition and the sense of adventure which must have driven the Brunelleschis of this world.

Brunelleschi was architect, engineer, constructor, inventor – each label would be appropriate in the context of today's specialized terminology. The idea of covering Florence's cathedral with a dome goes back to long before Brunelleschi came onto the scene. Building had started in 1296 and the decision to cover the cathedral with a single huge dome was taken in about the middle of the fourteenth century. But no one

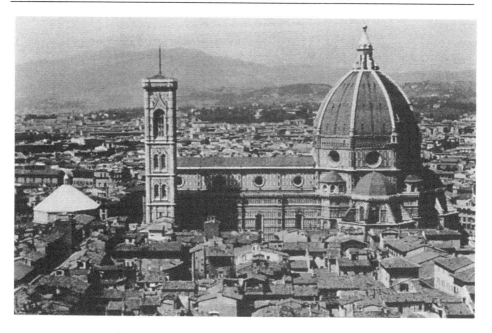

Figure 5.3 Santa Maria del Fiore. (Source: Steven Groak, Ove Arup & Partners.)

knew how to construct such a structure. There were no real precedents. The plan shape of the dome, its octagonal configuration and its height as well as the requirement for a massive surmounting drum were all fixed when Brunelleschi designed the dome itself, the lantern and the method of construction.

Even by today's standards the dimensions of the building are formidable. The dome diameter is about 42m, its height above the ground 84m and its rise 32m. It is sobering to consider the task of building this dome even with contemporary construction expertise. The skeleton of the cathedral existed, yet no one had any idea of how the enormous roof was going to be built. Available practical tools were rudimentary; the theoretical, particularly the analytical skills which would today be deemed to be prerequisite before even considering such an enterprise, were non-existent. Brunelleschi, along with others, competed to cover an existing building shell. His solution was daring beyond conception. In the face of some outlandish yet credible proposals by others, he not only devised the geometric proportions of the dome, but also a system of construction

which any engineer would be proud to design more than 500
years later. But that wasn't the end of it – he and his team
designed the machines, the pulleys and hoists necessary for
moving and hoisting the men and materials into position. Here
design and construction are in harmony. The innovative dou-
ble shell was a precursor for many others in the following
centuries. Yet with hindsight the question remains as to what
drove Brunelleschi to design a structure and its method of
construction against considerable odds and without what we
would consider to be rudimentary analytical reassurance that
his structure would not collapse.

Going further back in history to early Gothic structures, the
grandeur and awe inspiring spaces created by the great cathedrals in particular, are essential landmarks of our cultural life.
It has been suggested that Gothic buildings are arguably the
finest achievements in western architecture (Figs 5.4 & 5.5).
While the term 'great' can by no means be ascribed to all the
cathedrals of the times, there are common features which
support the argument.

Although the master builders of these cathedrals were generally masons in the language of the day, structural engineering no less than architecture can trace some of its genealogy
back to these monumental buildings. That is not to say that a
theory of structures as we know it today was available to their
designers, but they were able to organize the geometry of

Figure 5.4 Notre Dame. (Source: Peter Ross, Ove Arup & Partners.)

Figure 5.5 Norwich Cathedral. (Source: Peter Ross, Ove Arup & Partners.)

these buildings and they must have had a visual understanding of the distribution, if not of the magnitude, of the forces. Yet the stability of these daring structures could not safely be predicted. A kind of religious compulsion must have been present to spur these builders to such heights – literally and metaphorically! Dependence on precedence, on intuition and flair as well as experience gained from failures, often spectacular, was enough to fuel the ambitions, the aspirations and objectives not only of the builders but also of the promoters behind these great edifices. It has been suggested that Gothic builders set out to create pretentious self imposed structural problems. But that is to ignore the underlying theme which subordinates engineering means entirely to aesthetic aims. While these aesthetic aims included height and light from the great windows, yet – and being dangerously provocative – looking at these cathedrals centuries later, as structures, their logic can certainly be questioned. The eulogies accorded to the genius which produced so much of value to our civilization make any form of critical appraisal difficult and certainly unpopular, particularly when such appraisal is supported by hindsight made possible by the available theoretical and practical knowledge of the late twentieth century.

Gaudí, more of whom later, referred to the flying buttresses of Gothic buildings as unfortunate 'crutches' and set out to design structures which carried lateral thrusts to the ground more directly. If I can indulge in some more heresy for a moment, I could suggest that the great Gothic vaults are no more than decorated ceilings which unfortunately create large horizontal forces and that in order to maintain the stability of these extravagant features large buttresses, flying and otherwise, are necessary (Figs 5.6 and 5.7). Words to that effect might or might not have been uttered by some disgruntled medieval structural engineer! Yet the resulting structures have subsequently been revered and admired as glories of our civilization. The justification for these great buildings is man's spiritual and aesthetic aspirations rather than logic or economy. Without this kind of justification much of our civilization would not exist.

Moving now to more contemporary times, the Catalan architect/constructor Antonio Gaudí sought not only to rationalize traditional Gothic elements and images into the new technology of the late nineteenth and early twentieth centuries, but also to give it that extra spiritual quality which only the gifted architect/sculptor can create [7]. His buildings are an idiosyncratic synthesis of structural integrity and sculptural flamboyance which places him as one of the great designers and innovators of the last century. Again there are opinions which allege that Gaudí created structural problems to be able to express his virtuosity. However, there is no doubt that like

Figure 5.6
Notre Dame – flying buttresses. (Source: Peter Ross, Ove Arup & Partners.)

Figure 5.7
Norwich Cathedral – fan vault ceiling. (Source: Peter Ross, Ove Arup & Partners.)

his Gothic forebears, he subordinated structural means to architectural and spiritual ends. Gaudí's unfinished masterpiece, the church of Sagrada Familia in Barcelona is a testament to his inventive gifts (Fig. 5.8).

Gaudí rejected the fashionable Gothic revival of the late nineteenth century, a revival which attempted to replicate a 600 year old style without regard to contemporary technical expertise. But he used Gothic symbolism in a new and totally

Figure 5.8 Church of Sagrada Familia in Barcelona. (Source: Michael Bussell, © Ove Arup & Partners.)

personal and idiosyncratic manner. Gaudí's training and background was such as to enable him to develop his own technical solutions, often in refreshingly original ways. Most contemporary architects were beginning to leave this new vocabulary of structures to engineers. His knowledge of building techniques enabled him to embark on the experiments which became the basis of his novel structures, although an engineer collaborated with him for the numerical analysis. While in his more mature years the disposition of forces on his structures and their consequent expression became more and more overt, he nevertheless managed to create enough sculptural effects not only by the structures themselves, but also through decorative accretions which gave the whole a kind of romanticism (Fig. 5.9). In many ways Gaudí's work is a precursor to today's structures, many of which have become associated with the stylistic description of 'high-tech', a journalistic licence without historical perspective. Gothic architecture is higher than 'high-tech' and we must be careful not to become confused by subjective colloquialism. Gaudí has been described as an outsider – not so. In using structure to express his architectural objectives he not only did what many Gothic masterbuilders and Renaissance builders did before him, but also explored new forms which were intended to follow the thrust lines more directly.

Today the question as to what constitutes good architecture is possibly more open than ever before. Nowadays there are no rules. The plethora of architectural styles all have their protagonists as well as their detractors. Anything goes and the public is confused. Princely prejudice and intervention are less than helpful when an increasingly environmentally conscious society is trying to learn how to sustain and improve the built environment in all its aspects.

In the last 200 years or so, architecture has become the product of an ever-increasing list of specialists. While the role of the engineer varies from being the prime designer to, nowadays, more often a consultant, his contribution and more particularly the exercise of his technical skills are often seminal to the achievement of the architectural idea. How then should we respond when the architecture is new, strange or otherwise different from what we would normally expect? Are there criteria whereby we can sensibly make judgements so as to exercise our responsibilities to the public – and to our clients?

Whether or not some of the designers of the structures which have been referred to ever had doubts as to the validity

Figure 5.9 Sagrada Familia: sculptural effects. (Source: Michael Bussell, Ove Arup & Partners.)

of their enterprises, their efforts to explore new forms are interesting and open matters. It is a particularly interesting issue since the drive to enter unknown territory often originated with the promoters or at least with their active connivance, a luxury which is generally denied to us today.

Exceptionally, the Chief General Manager of the Hongkong and Shanghai Banking Corporation, Sir Vandeleur Grayburn, in 1929, asked an English architect practising in Hong Kong to 'please build us the best bank in the world'. This was the beginning of a tradition. The resulting building was one of the most innovative and completely equipped office buildings in the Far East or anywhere else for that matter. When the building, as a result of Hong Kong's increasing economic importance and the Bank's subsequent increasing fortunes, became inadequate, Sir Vandeleur's successors adhered to the quest for excellence. After a limited international competition held in 1979, they selected Foster Associates to be the architects for the project. The resulting building has become well-known for its architectural imagery, its capacity for change and growth, as well as for many innovative features in its design and construction (Fig. 5.10). It also features on Hong Kong's currency (Fig. 5.11). All these features are interesting but the central issue here is that again despite the apparent and to many, appealing technical imagery, structural means have rightly been subordinated to architectural ends.

If it is assumed that a column-free space at plaza level is necessary, there are a number of different means whereby the stack of office floors can be supported by the masts, the assemblies each of four columns, at each end of the building. Many options which were technically viable were examined, but only few met the rigorous functional and aesthetic criteria, so that the chosen solution correctly owes as much to functional need as to aesthetic preference. Moreover, at a tactical level vierendeel braces which connect the mast columns so that the four tubes act as a whole, had to be removed at plaza level (where from a purely structural engineering point of view they were most needed) so that people wouldn't walk into them (Fig. 5.12a) – a not unreasonable requirement! Removing these vierendeels added several hundred tons to the steel weight.

The main elevations which express the structure consist of the steel masts, the hangar trusses and the hangars. Like the flying buttresses and vaults of a Gothic cathedral they are part of the architectural imagery (Fig. 5.12b). However, the masts, trusses and hangars which form the elevations of the building carry little vertical load.

The resulting building has created international interest, admiration and some controversy (the latter mainly related to its cost which, when published, included complete fitout ex-

Figure 5.10 Hongkong and Shanghai Bank (Source: Ove Arup & Partners. Photograph: Ian Lambot.)

Figure 5.11 Hongkong and Shanghai Bank on $10 banknote.

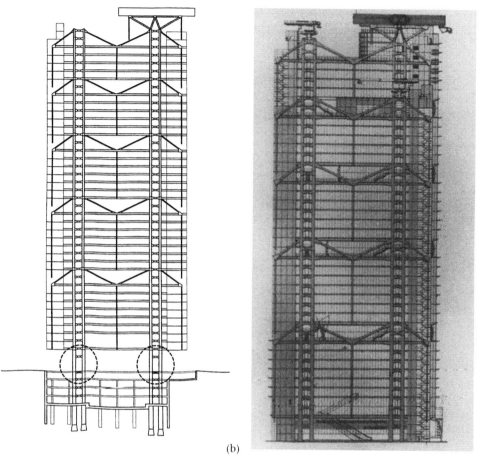

(b)

Figure 5.12 (a) Hongkong and Shanghai Bank: elevation of structure (© Hongkong and Shanghai Bank); (b) north elevation. (Source: Ove Arup & Partners.)

penditure not normally associated with the cost of the building shell and core). Despite the criticism the general consensus appeared to welcome the Hongkong Bank as a landmark in twentieth century architecture. The skeleton, the structure is seminal to its architectural expression.

Possibly one of the most important examples of twentieth century architecture is Sydney Opera House (Fig. 5.13). Its silhouette has not only become symbolic of Sydney's civic presence, but also as a landmark in what is most respected and admired in twentieth century urban design (Figs 5.14 and 5.15). It has become a symbol of Australia's emerging self-

Figure 5.13 Bird's eye view of Sydney Opera House. (Source: David Messent, Ove Arup & Partners.)

confidence. Recently it featured as the logo for Sydney's successful bid to host the Year 2000 Olympics. Yet before it was officially opened in 1973, there was considerable disquiet not only about its cost. There were many critics of its validity as a structure, as well as its value as a piece of architecture. Indeed there were many who questioned whether it should have been built at all.

Here is an interesting opinion:

'Architects . . . cannot escape this surrealist climate in which any outrageous gesture can produce worldwide although generally ephemeral, notoriety. Why descend to such prosaic details as satisfying oneself that a structure can be built? This task can be left to assistants of the second order, without any danger that such considerations might limit the creative capacity of the genius. The Sydney Opera House is

Figure 5.14 Water view of Sydney Opera House. (Source: David Messent, Ove Arup & Partners.)

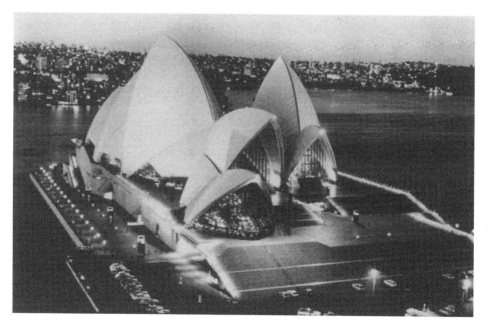

Figure 5.15 Close up of Sydney Opera House. (Source: David Messent, Ove Arup & Partners.)

a tragic example of the catastrophic consequences of this attitude of disdain for the most obvious laws of physics' [8].

Felix Candela
Arquitectura Mexico 1967

These words of Felix Candela, another Gold Medallist of this Institution, expressed deeply felt views about what he considered to be a monumental folly. He went to some lengths to exonerate Utzon, the architect who won the competition. He considered Utzon to be a man with considerable talent and he also spoke highly of the structural engineers. Nevertheless he talked about a 'catastrophe' as well as everyone concerned being carried away by 'an atmosphere of collective madness'.

Here then is one of the most respected designer/builders of the mid-twentieth century expressing in no uncertain terms his opinion on what was subsequently to be hailed as an architectural and technical masterpiece. This clearly underscores, if it needed underscoring, the dilemma in which we as structural engineers find ourselves. What is right, what is sensible, when are we wandering into or being led into the wilderness?

There are regrettably no simple answers. Indeed except in the case of some clearly immoral, or generally accepted anti-social objectives, our duty is to exercise our skills to the best of our abilities. There are, in our contemporary society, nihilistic and anarchic strands which are inevitably expressed in its music, its art and its architecture. Typically they are illustrated in the designs of Peter Eisenmann, one of the luminaries of contemporary architecture. A well-known architectural critic in an interview questioned him in the context of a house Eisenmann had designed for a client, a mathematician: '. . . part of you is proud of the fact that that mathematician couldn't live in the house and often you seem to be against content and against things working . . .'. His structural engineering colleagues are often faced with designing structures which are inefficient and defy gravity (Figs 5.16a and b). Paradoxically these seemingly random patterns more often than not are based on some clear, if intellectually sophisticated geometry, a geometry which has no meaning in practical construction terms but which is justified by its very special spatial impact.

I don't pretend to understand these buildings and I have questioned the sense of trying to build structures for what appear to be sophistic reasons. Nevertheless, some responsible observers consider the resulting buildings as signposts in contemporary architecture. Eisenmann's designs are often eulogized as much for the architecture itself as for the intellectual concepts behind them. Others are more critical, so that it is difficult to make a sensible judgement. Like so much modern art, to separate the real and lasting from the temporary and ephemeral divides even the most expert opinions.

Figure 5.16
(a) Nunotani building, San Francisco, upper level gallery;
(b) exterior. (Source: Guy Nordensson, Ove Arup & Partners.)

(a)

(b)

As engineers then our dilemma is a real one, particularly in the modern world where so much building design is architect led. We don't want to be party to designs which are against our consciences, nor should we be party to structures which we do not consider to be stable and appropriately durable in the context of all the knowledge and expertise available to us. While our historical predecessors could not justify their designs analytically, our newly found ability to carry out more and more rigorous analyses and tests should be a liberating rather than a constraining influence. When faced with problems which are new, different from our everyday experience, when we appear to be entering unknown territory we are better equipped than ever before. When working near the edge though, is it clear where to draw the line?

There *are* no further guidelines. Osbert Sitwell wrote that 'The artist, like the idiot or clown, sits on the edge of the world, and a push may send him over it'.

For artist read architect or designer. As structural engineers involved in the concepts of buildings, not only must we try and stop the architect from going over the edge (and probably taking us with him) but also and paradoxically make sure that he does from time to time get to the edge in the first place. In an increasingly confrontational, litigious and anti-technological atmosphere it is only too easy to adhere to the tenets of conventional wisdom. 'Value for money' is the slogan which has come to mean the lowest possible price one can get away with regardless of quality and real cost, and particularly aesthetic satisfaction.

We must vigorously resist this climate which is a fertile breeding ground for defensive professional attitudes. We must continue to take the lead in creating new structural systems, explore new materials and their forms and designs at the boundary of our technology. But if we do look at the stars from time to time we must also keep our feet firmly on the ground. Pushing back the frontiers is our duty if we are to honour the great engineers of the past and at the same time give guidance to those who follow us.

References

[1] Kerensky, O.A., An engineer's ethics. *The Structural Engineer*, December 1970.

[2] Bennion, F.A.R., *Professional Ethics*, C. Knight & Co. Ltd, 1969.

[3] Lord (R.A.) Butler, The Professional Man in Society, Gold

Medal Address, 12 June 1968, Royal Institution of Chartered Surveyors.

[4] Mordaunt Crook, J., *The King's Works*, Vol VI, 1782–1851, H.M. Stationery Office, 1973.

[5] Battisti, E., *Brunelleschi – The Complete Work*, (trans. R.E. Wolf), Thames & Hudson, 1981.

[6] Parsons, W.B., *Engineers and Engineering in the Renaissance*, The M.I.T. Press, Massachusetts.

[7] Sweeny, J.J. and Sert, J.L., *Antonio Gaudi*, The Architectural Press, London.

[8] Candela, F., *Arquitectura Mexico*, **98**, 1967, 103.

6
History of structural design

Frank Newby

P.J. Barnett

Frank Newby studied at Trinity College, Cambridge from 1944 to 1947. He joined Felix J. Samuely in 1949 as a junior engineer and later became research assistant. Awarded a US Government scholarship in 1952 to study building methods in the USA he worked with Wachsmann, Eames, Saarinen and Severud. In 1953 he returned to Felix J. Samuely and became a partner in 1956 and senior partner on the death of F.J. Samuely in 1959. He retired in 1991.

Frank Newby has been responsible as structural engineer for many notable buildings in this country and in Saudi Arabia. He is keenly interested in education, lectures widely and has recently taken over from James Sutherland as convenor of the I.Struct.E. History Study Group.

At the symposium on the relevance of history at this Institution in January 1975, Sir Alfred Pugsley stated that 'each generation does indeed advance by standing on the shoulders of its predecessors' thus echoing Sir Isaac Newton's observation, 'If I have seen further it is by standing on the shoulders of giants'.

Felix Samuely, a structural engineer, reached London in 1933 after practising in Berlin. He was engaged in the design of the Bexhill Pavilion, the first welded-steel frame public building with architects, Mendelsohn and Chermayeff and later became the engineer most closely associated with the Modern Movement architects in this country. He was also a brilliant mathematician which added substance to his innovative ideas. In 1949 I was lucky enough to be directed towards Samuely for work and in time I began to realise what a giant I had chosen.

During the Second World War Samuely started teaching structure at the Architectural Association in London and afterwards set up his office again. The lack of building materials and skilled craftsmen in the immediate post-war period led to the

Structural Engineering: History and development, Edited by R.J.W. Milne. Published in 1997 by E & FN Spon, London. ISBN 0 419 20170 X.

use of prestressed concrete with its minimum weight of tensile reinforcement, and of factory produced components. Samuely's first use of prestressed concrete in a factory in Bristol was perhaps his best, with heavily-loaded floors being made up of storey-height precast concrete frames with prestressed concrete ties. In other areas of the building he used pretensioned concrete planks as reinforcing bars in continuous slab construction. At the Festival of Britain his welded tube and rod latticed trusses gave a feeling of lightness to the Industry and Transport Pavilions, while for the Skylon he streamlined the structure by prestressing the cables and by keeping all cable anchorages out of sight.

I became Samuely's assistant and worked with him on his initial design work with architects, meeting his clients and becoming his back-up link. I was lucky to be around during this fascinating and innovative period. As with the early railway period a certain amount of risk was taken by society at this time, not only with a new housing philosophy which included multi-storey tower blocks, but also with novel construction techniques. Samuely prestressed concrete, steel, timber and brick. He introduced me to star beams and to folded-plate roofs which illustrated his mastery of thinking of structure in three dimensions.

I accepted a partnership in 1956 after spending a year in the USA in 1952–53 studying the building industry and took over as senior partner at the age of 32 on Samuely's premature death in 1959. By then I had been in charge of the design of the US Embassy in London and the British buildings at the Brussels Exhibition. I had also travelled widely and had seen most of Nervi's buildings. I also lectured and advised architectural students on their schemes and in time I realized that this was excellent training for an engineer who wanted to work closely with architects. My experience could not therefore have been better, particularly with regard to the conception of structure which is the aspect of structural design on which I want to concentrate.

In June 1962 I gave my first major lecture at the Architectural Association [1]. I produced for it a 'decision tree' (Fig. 6.1) that I felt structural engineers followed in the course of their work with architects. At that time I was interested in the nature of initial design (Fig. 6.2) and how the engineer conceived possible structural solutions. I came to the conclusion that there were relatively few innovative engineers and that experience was the most important source of inspiration.

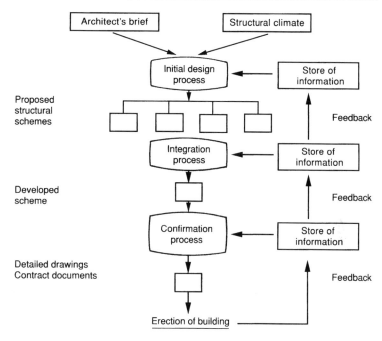

Figure 6.1 Decision tree.

As the function of structure is to carry loads safely into the ground, I introduced the idea of 'routes of stiffness' along which the loads flow in varying amounts depending on the relative stiffness of those routes. I suggested that engineers or architects consciously or subconsciously conceive structure in terms of deformation or strain as against stress and that they are free to position these routes of stiffness anywhere in space; their positioning and sizing is an art. However, competence as a structural engineer with an ability to create structural form does not imply capacity to create architecture.

On the other hand, not many architects have the understanding of technology necessary for design in the present, increasingly industrialized building climate, and collaboration between architect and engineer is therefore essential. The consideration and analysis of possible structural solutions, I suggested, was carried out in the confirmation process for which engineers are trained, for the engineer's responsibility to society is to make certain that structures do not collapse, an operation concerned with stress.

But a structure has to fulfill many architectural functions in

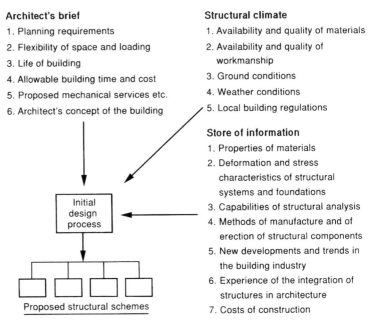

Architect's brief

1. Planning requirements
2. Flexibility of space and loading
3. Life of building
4. Allowable building time and cost
5. Proposed mechanical services etc.
6. Architect's concept of the building

Structural climate

1. Availability and quality of materials
2. Availability and quality of workmanship
3. Ground conditions
4. Weather conditions
5. Local building regulations

Store of information

1. Properties of materials
2. Deformation and stress characteristics of structural systems and foundations
3. Capabilities of structural analysis
4. Methods of manufacture and of erection of structural components
5. New developments and trends in the building industry
6. Experience of the integration of structures in architecture
7. Costs of construction

Initial design process

Proposed structural schemes

Figure 6.2 Initial design process and structural solutions.

a building besides carrying loads safely into the ground. At present it is the architect who is trained in the history and conception of building including its structure and it is he who has a moral responsibility to society for its appearance. He uses his consultants as a sculptor uses his tools and as such he looks for those who can provide innovative ideas and constructive information on the latest technology.

So far I have not mentioned history directly. What I have termed information, on which the conception of structural form is based, contains a great deal of the history of the immediate past. How can those, who have not had giants' shoulders to stand on, be informed of the developments of the immediate past and who is to guide, stimulate and criticize their structural inventiveness? This cannot be done at university unless there is a radical change in the attitude of the staff in teaching theory and the use of part-time practising engineers to teach conceptual design. Perhaps a better scheme would be to have a year's post-graduate course for all professionals going into the building field where the role of structure in architecture, the impact of computer technology on structural analysis and the history of the immediate past would be three areas for study and research. In time the independent structural engineer could well become a specialized member of a comprehensive building design group and might even take over architectural design responsibilities.

Another, quite separate field of building where the engineer has an important contribution to make is in the refurbishment of historic structures. For this he must have an enquiring and open mind on how loads are being transferred into the ground. The demolition of any one route of stiffness could cause overstressing and, in the worst case, collapse of other routes. The assessment of the adequacy of an existing structure also needs some understanding of the technology at the time of construction and of the history of the use of the building. This task is quite different from designing a building on a green-field site guided by present day building regulations and is one for which he is not trained or even introduced to at university. I believe that the Institution has a responsibility to cover this omission and should promote discussions on refurbishment and circulate relevant information.

Let us now look at the prehistory of the immediate past where I would like to discuss some engineers who made outstanding innovations in the conception and understanding of structure.

J.R. Perronet, the father of French civil engineering, was the first engineer to alter the appearance and economy of multi-arched bridges. Medieval bridges were built arch by arch with the piers sized to take the thrust from the individual arch. In 1757 Perronet noticed how a pier moved sideways but returned to the vertical as adjacent arches were built. He realized that if he built the whole bridge in one operation, removing all the centring at the same time, the end buttresses would take the thrust leaving the piers carrying only vertical load. They could therefore be much smaller than the piers of traditional bridges and would cause less interference to the flow of the river. With substantial buttresses in good ground, very high thrusts could be taken which meant also that the arches could be shallower. Perronet effectively changed the span-to-depth ratio from 3 to 11.

The first man-made material to be used on a large scale was cast iron. In 1796, within two decades of the construction of Ironbridge, the engineer Thomas Wilson clearly expressed the characteristics of this material in his 236ft span single arched bridge at Sunderland. It was made up from identical cast-iron voussoirs bolted together and tied laterally. Because of the similarity between cast-iron and pre-cast concrete I wrote an article [2] comparing the two in which I made the point that in each case structural material was poured into moulds and that their shaping or sculpting was in the hands of the designer. With the present development of SG or ductile iron, a survey of cast-iron techniques would now be relevant.

Thomas Telford, with training and experience as a stonemason and architect, became a road, canal and bridge builder. His structures of both stone and iron are masterpieces of architecture and engineering. In 1800 he was sufficiently confident in his handling of cast-iron to propose a cast-iron arched bridge as a replacement for the medieval London Bridge with a single huge span of 600ft, using the same technique as at Sunderland. The scheme aroused considerable interest and the best engineers, ironfounders and scientists of the day were asked for their comments. These were included in an outstanding Government report of June 1801 [3] of which Telford remarked that it 'would be the means of throwing much new light on this important subject [cast iron] and will probably change the principles and practise of this species of architecture'.

I.K. Brunel needs no introduction as one of the giants of the railway age, notable in particular for his bridge at Saltash and his 'Great Eastern' steamship. His prowess with timber trusses, masonry arches and wrought-iron tubes is well described in the book edited by Sir Alfred Pugsley [4] but I would like to describe his conceptual design for the proposed suspension bridge at Clifton in Bristol.

In 1828 Brunel was recovering in Bristol after an accident in the Thames Tunnel, designed by his father, M.I. Brunel, when it was proposed to hold a competition for the design of a bridge across the Avon Gorge. His father was asked to compete but passed the invitation on to his son, then 22 years of age, who promptly inspected existing suspension bridges in this country and discussed their mathematical design with Davies Gilbert, the leading authority. He also made a detailed survey of the Gorge to establish possible sitings of the bridge. The Gorge varies in width from 700ft to 1000ft: the maximum span of a suspension bridge at that time was the 580ft of Telford's Menai Bridge.

Brunel put forward four different schemes for this, his first major project. He considered the rock at either side of the Gorge capable of anchoring the suspension chains and some of his designs had spans of up to 900ft from rock to rock with the roadway approaching the Gorge through tunnels. Another design had towers on either side of the Gorge which reduced the span to about 650ft.

In his submission to the Committee he talked first as an architect about the majesty of the location and how a bridge must enhance it. He remarked of his ambitious designs that engineers must work at the cutting edge of technology otherwise their successors would criticize them for their waywardness. He produced marvellous drawings of his proposals which can be seen in his notebooks, which are now held at Bristol University Library. They are a delight to view and show his brilliant faculty for illustrating his designs on paper.

Robert Stephenson was another famous railway engineer. He was confronted in 1845 with the problem of how to cross the Menai Straits with a bridge of four spans, the maximum span being 460ft (the previous largest span for a railway bridge was 150ft), able to carry the heavy moving load of a locomotive. The area beneath the bridge had to give clear access for shipping so cast-iron arches were not possible.

Initially Stephenson proposed the bare outlines of a tube through which the train would run and when the necessary Parliamentary Bill was passed, he was left with the problem of what size and shape the tube should be and how to build and erect it. The technique of riveting wrought-iron plates in ships to produce 'tubular' construction had been developed by the ironmaster, William Fairbairn, and he was appointed to assist Stephenson. They started with model tests on tubes with a circular cross section, then an oval section, then one with a stiffened top and eventually a rectangular section, first with small round tubes as the upper chord and then built-up rectangular cells top and bottom. For the final shape, a sixth full-size model was tested at Fairbairn's shipyard at Millwall. In order to confirm test results, Eaton Hodgkinson, an eminent mathematician and experimenter, was brought in by Fairbairn to carry out a structural analysis though the final calculations appear to have been completed after the construction of the tubes had begun. Fairbairn wrote a book on the bridge in 1849 [5] followed by that of Edwin Clark, the resident engineer, a year later [6]. Both books, coupled with the collection of related correspondence held by the Institution of Civil Engineers, show the exchange of ideas between Stephenson and his associates in bringing the initial concept to reality.

The impact of the use of mathematics to check the strength of the wrought-iron tubes for the Britannia Bridge was considerable. Until that time individual cast- and wrought-iron beams were test loaded at the works before delivery to site and all railway bridges were test loaded on site after completion. Using calculations it became possible to prepare tables of the sizes and strength of beams, and later of trusses. Testing became outdated and innovation began to disappear as standard calculated beams became the norm. It has also been suggested that this was the time when architects and engineers parted company and went their own ways.

J.B. Eads, a mechanical engineer, holds a special fascination for he started with a steam boat on the Mississippi at St. Louis and knew the bottom of the river intimately from his underwater salvage work. In 1867 he proposed at St. Louis a three-span latticed arched bridge with a central span of 500ft to carry both road and rail.

In the presentation of his scheme, he made some telling remarks:

> 'If there were no engineering precedent for 500ft spans can it be possible that our knowledge of the science of engineering is so limited as to teach us whether such plans are safe and practicable? Must we admit that because a thing has never been done, it can never be, when our knowledge and judgment assure us that it is entirely practical? This shallow reasoning would have defeated the laying of the Atlantic cable, the spanning of the Menai Straits . . . and left the terrors of the Eddystone without their warning light!' [7]

For the tubular chord members Eads used cast chrome steel, a new material recently used for a bridge at Coblenz which he had seen. He appreciated the potential saving in dead weight by using steel at a higher design stress than that of wrought-iron and with a fresh mind came up with a design of great inventiveness. The tubes, 9ft in length, were made up from 10 segmental bars restrained by a 1/8th inch thick cover plate similar to the hooped construction of a barrel. The bars were machined at their ends and fitted into a coupler which not only provided a change in direction but by a single pin connected them to the latticed diagonal. This design could only have come from the hand of a mechanical engineer steeped in steam engine technology.

Maillart and Freyssinet are the two early giants of reinforced concrete. Robert Maillart built his first unreinforced arch bridge in 1900 before visiting the Paris Exhibition where reinforced concrete for building was first demonstrated on a major scale. He turned to reinforced concrete and, acting as consultant and contractor, he refined and developed his arched bridge design, finally producing the Tavanasa Bridge in 1905, clearly recognizable as the classic Maillart three-hinged frame bridge. The next development was the use of a thin arch slab loaded by cross walls with heavy, stiff parapet beams which distributed concentrated applied loads onto the arch. For the Klosters Railway Bridge and later for the Schwanbach Bridge, he used the same principle but curved the deck on plan. He combined the inherent stiffness of the arch, the vertical walls, the deck slab and the parapet beams to produce a three-dimensional form, a true expression of *in-situ* reinforced concrete. The form did not originate from calculation but from Maillart's experience and intuition.

In 1938 he wrote:

> 'Reinforced concrete does not grow like wood, it is not rolled like steel and has no joints as masonry. It is more easily compared with cast iron as a material cast in forms . . . The engineer should then free himself from the forms dictated by the tradition of the older building materials so that in complete freedom and by conceiving the problem as a whole it would be possible to use the material to its ultimate'. [8]

Eugène Freyssinet is probably the most important giant of this century and Sir Alan Harris was privileged to have stood on *his* shoulders. Freyssinet's earliest experience was on the construction of large-span reinforced concrete arched bridges. His innovative ideas included the use of hydraulic jacks at the springings for decentering. He also used similar jacks at the crown of his Veudre Bridge to relevel it after it had settled by 5in. due to creep. In 1918 he joined a contractor to design both bridges and buildings. His Orly hangars of 1918–21 were made up of reinforced concrete arches of 295ft span and 190ft in height using 3in. thick corrugated slabs. He went on to build thin shell roofs which have seldom been illustrated.

However, he became famous for his innovative idea for prestressing concrete. He realized that he needed both high-strength concrete and high-tensile steel so that sufficient precompression remained in the concrete after creep and shrinkage movements. In 1928 he was able to take out a patent and to start working on his own.

By 1931 he was producing prestressed concrete lamp posts in a special factory having solved all the technical problems of high-strength concrete, but there was no market and he lost a fortune. In 1934 his luck changed when he saved the Marine Terminal at Le Havre from collapse (it was settling at the rate of 1in. per month) by adding extra concrete between existing foundations and by prestressing them together to form a continuous grillage which transferred all the loads to a new foundation of prestressed concrete piles. From that time onwards the idea of prestressing was accepted and in 1936 his first bridge, spanning 64ft, was built.

He wrote,

> 'A system of prestressing is not limited to a predetermined technical field. It is, in reality, a state of mind, an affirmation of the will of the engineer to accept no longer the consequences of the initial elastic states brought about at random by the method of construction. On the contrary, using prestressing as one more datum of his projects, the engineer can modify this state at will, just as if dealing with the resistance of the beams.'

Freyssinet, unlike Maillart, was an engineer's engineer, seeking for truth in the behaviour of materials and structures and lacking interest in art and architecture. However, his structures and his monumental invention of prestressing have inspired twentieth century engineers. On the subject of conception of structure he made the following comments,

> 'There are two possible sources of obtaining information: one is a direct perception of reality, the other is intuition. By this I mean the sum total of human experience built up over the years in the subconscious.'

and

> 'It is only natural that intuition should be controlled in the light of experience, but when it turns out to be in direct opposition to some calculated result I always make a double check on the calculation . . . in the long run, it is almost always the answer determined by calculations that turns out to be wrong.' [9]

Pier Luigi Nervi, the leading architect's engineer, was a giant who, fortunately for us, was the author of a number of books in which he clearly reveals his attitude towards his material, concrete. In one of them, *Aesthetics and Technology in Building*, he devotes a significant amount of space to the history of construction before describing the design concepts of his major structures. He completed his first major project 60 years ago, a stadium in Florence and the staircases were his best expression of the sculptural quality of concrete. He later developed ferro-cimento and with his isostatic slabs moved out of the rigid grid mentality. His domes have not been surpassed in beauty, and I caused consternation at my 1962 lecture by describing them as jewels in an indifferent setting designed by others.

Torroja, Candela, Buckminster Fuller, Frei Otto, Samuely, Arup, Fazlur Kahn and Peter Rice are some of the innovative giants who have followed Nervi in the field of building and who have made significant contributions to our rich history of structural design.

For an exhibition on the art of structures in 1982 [10] I included illustrations of the work of some of the giants mentioned. In the introduction to the catalogue I suggested that three questions should be asked: why, where and what is a structure? The first two were reasonably straightforward to answer. I stated that 'advances in the art of engineering have usually come about for specific reasons. The railway era created new problems in scale and loading of its structures as do the offshore drilling rigs today. War, too, poses new problems solved, for example, by inflatable bridges, Mulberry harbours and supersonic aircraft'.

Structure I defined as 'natural or man-made bits interconnected in space. Its essence and art lies in the understanding of how these bits work individually and together and in joining them in such a way as to form a stiff and stable whole. There is no science or law to say how the bits should be distributed in space. That is a decision made entirely by man, the engineer. To conceive a structure some knowledge of or instinct for stability is essential and this may come from an awareness of three-dimensional forms, from trees to primitive tents to skyscrapers. Stability is three-dimensional. Loads are transmitted into the ground via various structural "routes", each having its own stiffness. The stiffer the route the more load is attracted and the distribution of stiffness in space is the art of the engineer.'

Since 1982, there has been a marked increase in the capability and use of computers. Possible systems can be checked quickly and more complex forms can be investigated. But it is nevertheless still necessary to be trained in the conception of structures. Out of interest for an article on my work in 1985

[11], I described my major structures and the influences which led me to each design solution. Certainly the work of Samuely, Nervi, Fuller and Wachsmann among the designers of the immediate past was a prime source but in other cases it was the architect who sparked off an idea which I was able to develop into a new and exciting form because of my wider appreciation of the nature of structure brought about by a study of the history of engineering. Such a study provides the engineer with a sense of presence in the continuous evolution of the construction industry. Without an appreciation of the past how can the present be assessed and the future foreseen?

James Sutherland, in his inaugural Sutherland History Lecture concluded his observations on the use of history as an aid to teaching engineers in greater depth thus:

'My aim would be to provide a path into structural engineering for more students with a broad imagination and a visual, historical bent who can draw and write and wish to develop designs rather than solve set problems. At present these people tend to go into architecture or industrial design. We need more of them in our profession and I see no better base for educating such people than engineering history'. [12]

For civil engineering there is one further aspect to be considered. Unlike the structural engineer, the civil engineer is the lead consultant and thus has a moral responsibility to society with regard to the appearance of his projects yet he has no training in conceptual design or in aesthetics. It is surprising therefore that the institutions do not require any ability in or appreciation of visual design for their professional members who are entrusted with these commissions.

Let me quote Inglis in 1944 when he remarked, in a lecture at the ICE on bridges, 'A striving after beauty of form and harmony with surroundings is a social obligation which structural engineers must recognise and educate themselves to perform' [13].

The recent impact of Santiago Calatrava, sculptor/architect/engineer, with his exciting and innovative bridge designs is very revealing. Although he does not use the most economic forms, he nevertheless commands the attention of the public as did Stephenson and Brunel. Only a rethink on these major issues of training, in which the study of history and aesthetics will play a vital part, can restore to the profession the full respect of society.

References

[1] Newby, F. (1962) Architect and engineer. *Architectural Association Journal*, **78**.

[2] Newby, F. (1960) On cast iron and precast concrete. *Architectural Design*. August.

[3] *Ibid.*

[4] Pugsley, A. (ed) (1976) *The Works of Isambard Kingdom Brunel.* London, Inst.C.E.

[5] Fairbairn, W. (1849) *An Account of the Construction of the Britannia and Conway Tubular Bridges.* John Weale, London.

[6] Clark, E. (1850) *The Britannia and Conway Tubular Bridges.* London, for the Author.

[7] Woodward, C.M. (1881) *A History of the St.Louis Bridge.* Jones & Co., St.Louis.

[8] Bill, M. (1949) *Robert Maillart.* Verlage für Architektur, Erlenbach-Zürich.

[9] Fernandez Ordoñez, J.A. (1979) *Eugène Freyssinet,* 2C Éditions, Paris.

[10] Newby, F. and Elton, J. (1982) *The Engineers.* (Catalogue of an exhibition held at the Architectural Association). Architectural Association, London.

[11] Newby, F. (1985) Perfectly Frank. *Building Design,* 31 May.

[12] Sutherland, J. (1994) Active engineering history. Paper given at the I.Struct.E., March 10th.

[13] Inglis, C.E. (1945) The aesthetic aspect of civil engineering design. Third lecture, 3 May, London, Inst.C.E. 1945.

7
Design for the developing world

Jörg Schlaich

oto Hostrup-Zehnder

Jörg Schlaich is professor for structural design at the University of Stuttgart and consulting engineer with Schlaich Bergermann und Partner, Stuttgart. He has designed bridges, especially footbridges, towers and long-span cable-net-roof, membrane- and glass structures as well as solar energy plants. He has also written several papers and books on structural concrete research and design.

It is more than a challenge, rather an impertinence for a continental European – a German even – to accept this topic for a lecture in the capital of Great Britain. What we now choose to call 'developing countries' is by history and practice so much more familiar and closer to a Briton than to a German, that it would have been more adequate to have this subject presented by a British Gold Medallist. Well, since on the other hand you consider the willingness to always follow orders, even unreasonable ones, to be one of those 'lovable' German characteristics, I will face this task put to me by the President without any further complaints. Anyway, I will not be so presumptuous as to cover it universally and comprehensively, but shall limit myself to describe some minor subjective experiences with this matter. A personal 'ailment' allowed me to collect these experiences or impressions, the travel-bug.

In 1976 the whole of my family followed the 'British Trail' in a VW van: Europe, Turkey, Persia, Afghanistan, Pakistan and all the way through India, to Calcutta, because there we were doing in the true sense a 'Design for the developing world', the Second Hooghly Bridge, at that time the largest cable-stayed bridge in the world with a main span of 457m and 35m wide. It was to be designed towards a completely indigenous manufacture and erection. Steel was to be preferred as against concrete in view of Calcutta being a steel city with several large

Structural Engineering: History and development, Edited by R.J.W. Milne. Published in 1997 by E & FN Spon, London. ISBN 0 419 20170 X.

workshops rich in tradition. It was to be rivetted throughout, without or with a minimum of welding. 'We', that were first of all British consultants, Rendell Palmer and Tritton, who initially had proposed an arch bridge, and Freeman Fox and Partners' Bernard Wex, Bill Brown and Mike Parsons who under the leadership of the great Oleg Kerenski had prepared an alternative design for the contractors Braithwait, Burns, Jessops and Gammon, India, a cable stayed bridge. In view of the fact that the two main longitudinal girders of this initial design were box girders, which at that time had caused some irritations, the client, the Hooghly Bridge Commissioners, looked for an additional consultant and invited several for qualification. As a junior partner of Professor Leonhardt, I happened to be in India at that time. Leonhardt knew of my interest in this country and therefore accepted the invitation of an Indian contractor to help him with the design of the 5km long cable stayed concrete bridge across the Ganga at Patna, though he foresaw from earlier experience, the Allahabad Bridge, that the chance to convince the Indian authorities to go for such a novel design was nil. So I passed by in Calcutta in 1971 and Leonhardt and Andrä got the job. What a job! It took about 20 years to complete it and changed many lives (Fig. 7.1). Doing the site supervision, we not only supplied active membership to the Tollygunge golf club but also brought home several Indian wives. At the beginning it was not that easy, in fact, it was never easy: ask my partner Rudolf Bergermann. We had to agree on a design revision omitting the box girders entirely, in favour of open I-sections forming a grid (Fig. 7.2). This grid was simple to manufacture and could be erected easily in the free cantilevering method, typical for cable stayed bridges. Thereafter, it served as a permanent falsework for the subsequent concrete slab, connected with it by shear-studs and relieving it from the compression forces which the concrete gratefully accepts. The two cable-planes spaced 29.1m apart deliver their forces directly into the webs of the main girders, without any special cross-girders there. Of course, now the cables have to be stressed from the pylon-heads, which in fact can be done there much more conveniently in one chamber, rather than moving along the main girder as usual. The pylon-heads respond with a characteristic shape nick-named 'flower pots'. The whole bridge was manufactured and erected by the Indian contractors with immense precision, care and dedication and without any major drawback.

At the time of the design of this bridge, we all felt that this

(a)

Figure 7.1 Hooghly River Bridge from above; (b) view from the side of the bridge. © Prolab GmbH.

(b)

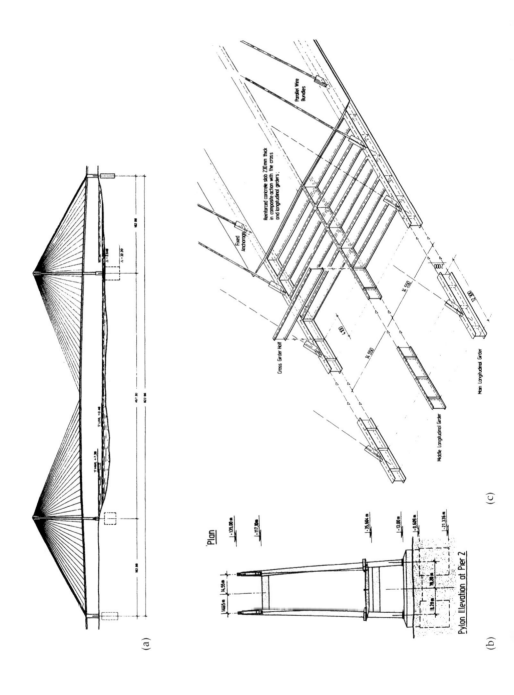

(a)

(b)

(c)

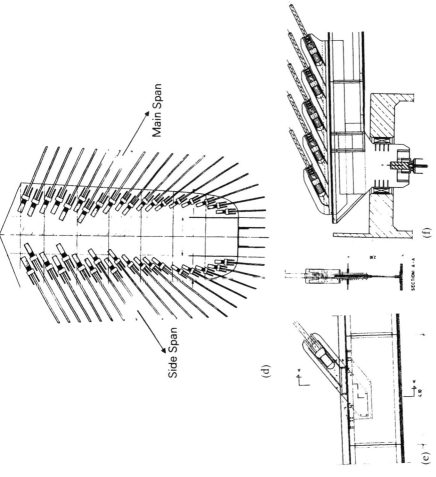

Main Span

Side Span

(d)

SECTION A-A

(e)

(f)

Figure 7.2 Hooghly River Bridge, Ca cutta, India: (a) view; (b) section at pylons; (c) the composite deck: steel grid with concrete slab; (d) mast top; (e) cable anchorages at deck; (f) anchorage and holding down of back-stay cables combined with the horizontal bridge support.

is a 'good enough' solution for a developing country, avoiding
any sophisticated devices for welding an orthotropic deck,
then the international standard for such cases. On the other
hand, a pure concrete deck was to be excluded, because of its
weight and its impact on the cable costs. So the composite deck
appeared to be the right compromise here.

Today, and at least since the very successful construction of
the Annacis Bridge in Vancouver, Canada, we know that a
composite deck is the natural solution for a long-span cable
stayed bridge in industrialized countries also. It evolved, however, on a detour from the satisfaction of 'typical developing
world boundary conditions': simple, efficient, robust, unso-
phisticated, indigenous, sustainable. Why?

Because during the planning and construction period of this
bridge, our world changed fundamentally from technology –
to being environmentally – or even ecology-minded, and these
became exactly the values we care for today. So this bridge,
beyond introducing us into a new cultural universe and letting
us make lifelong friends across the continents, has also taught
and given back professionally more than it got from us: new
values. Last but not least, it rewards the spectator, layman or
engineer, with its clean, simple but nevertheless – due to its
sheer size – majestic appearance, where obviously nothing
can be left out and nothing must be added. India must be
proud of it.

Of course, in general, in developing countries, concrete is
the cheaper and more adequate material than steel. At the
same time, for countries like India, where the streams swell to
huge rivers during the monsoon with several kilometres
width, cable stayed bridges are the right solution (Fig. 7.3).
Whereas a prestressed concrete box girder may span not more
than 100m, a concrete cable stayed bridge conveniently man-
ages 300m and more, and can be added to any length, thus
reducing the main cost factor, the number of wells and piers,
to at least 1/3. Due to the immense scour depth of such rivers,
the wells must be sunk to say 70m into the river bed and thus
reach diameters of 15 to 20m. These huge wells can easily take
the loads of a 300–500m cable stayed bridge, even if completely made from concrete without any additional costs. After
unsuccessfully trying to convince the Indian Ministry of Transport for a decade to build such a bridge, we happily accepted
the invitation of Dr T.N. Subba Rao, then managing director of
Gammon, India, to build at least a small prototype, a 2 × 70m
cable stayed concrete bridge in a remote Himalaya valley, the

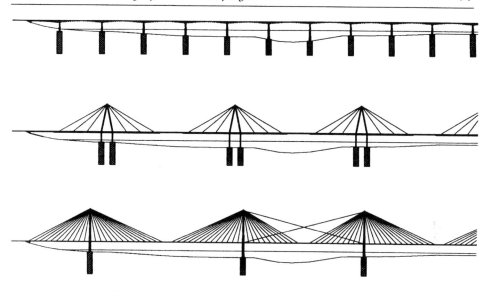

Figure 7.3 Girder bridges versus multiple span, cable stayed bridges crossing rivers requiring deep wells.

Akkar Bridge in Sikkim (Fig. 7.4). Without trying to explain this bridge in detail – because we shall find all its features again, although improved, in the Evripos Bridge in just a moment – I was most impressed how it could be done locally with simple though up-to-date means: modern technology must not be sophisticated or elitist. It should serve people always in two ways simultaneously: not only the result counts, here the bridge carrying traffic across a hindrance, but also the jobs it provides, the chance for people to make a living with their hands, and the satisfaction this gives. Lean production, I am sorry to say, is inhumane. It cannot be the result of a holistic thinking. Producing ever more goods, ever faster with ever more robots supervised by ever less people, the rest sitting idle, with technology becoming an end in itself, is a dead end. It especially leaves the Third World, three quarters of humanity ever further behind, because its only capital is its cheap labour. So I am proud to say, that for the Akkar Bridge we even produced the cables on site. We developed a special parallel wire cable made from local prestressing wire. It has a long lay to keep its hexagonal shape and is filled inside and covered outside with 'paint' and anchored in cylindrical sockets with conventional zinc cast. Some of my university collaborators, who had developed and tested these cables – for

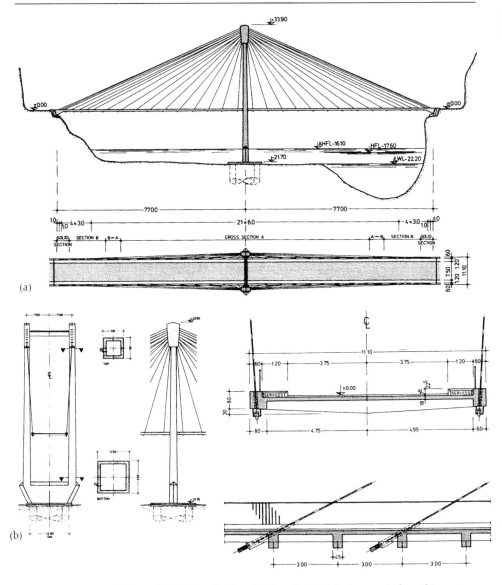

Figure 7.4 Akkar Bridge, Sikkim, India: (a) view and plan; (b) mast; (c) girder with cable anchorages.

example Knut Gabriel – went to the site to introduce these cables to the Indian workers and everybody gained, the bridge and all those involved.

Even the Evripos Bridge in Greece gained (Figs 7.5 and 7.6). It is another example where a design towards simplicity – we

Figure 7.5 (a) view of the Evripos Bridge; (b) during construction. © Prolab GmbH.

did it together with our Greek colleague Dr S. Stathopulos – made in view of developing world boundary conditions, paid off in a developed country (whatever this means). In this case though, it was also for Greece the first cable stayed bridge to be built by a local contractor who had in fact much less bridge experience overall compared to his Indian colleagues, who by size and number of the bridges they built are unmatched.

With his Diepoldsau-Bridge, René Walther had shown that for medium span and small width the solid concrete slab

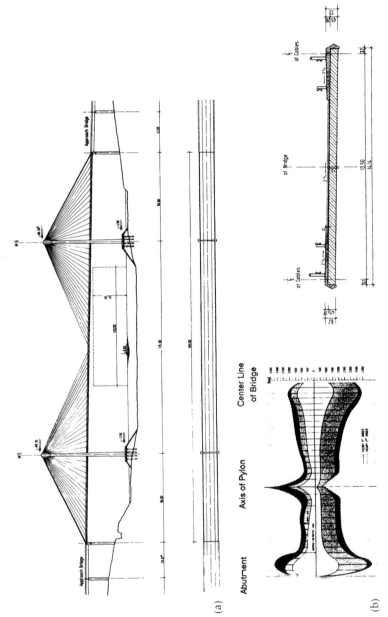

Figure 7.6 Evripos Bridge, Greece (a) view and plan; (b) section through deck slab and moment envelopes due to Lve load.

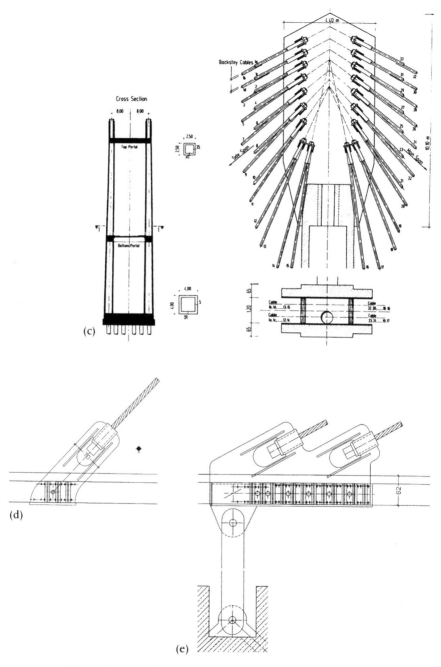

Figure 7.6 *cont'd* (c) pylons and mast top cable anchorage; (d) cable anchorages at deck level; (e) back stay anchorage with holding down pendulum.

makes an optimum cable stayed bridge girder, simple and robust. The deck of the Evripos Bridge with a main span of 215m and two side spans of 90m each is 14.14m wide only and therefore suitable for a solid concrete slab with a constant thickness, 45cm in this case. It is monolithically connected to the concrete pylons, since these are high enough between the foundation and the deck level to respond elastically to the temperature expansion of the deck. Only at the ends are movable supports provided. On the other hand, thanks to the small bending stiffness of the slab, the moments which build up at those vertically rigid supports, as against the spring supports at the cables, are not excessive but of the same order as those due to live loads at other sections. The benefits of such a monolithic connection are an increased overall robustness, including an improved ductility with respect to seismic attack, and savings in bearings and in cable steel, because the deck loads enter the pylon base directly and avoid the detour through the cables and the top part of the pylons.

The method of cable anchoring and stressing basically followed the approach used for the Hooghly Bridge. In designing the pylon heads for a smooth and direct flow of forces and to avoid any prestress or sophisticated reinforcement a stressing chamber was provided with its inner faces lined with steel plates. The pairs of cross-bars, provided to support one cable socket each, transfer their loads to these plates in which the horizontal components of the cable forces can easily balance in tension whereas the merging vertical components are transferred to the concrete through shear-studs. This results in a very compact cable arrangement, close to the ideal fan-geometry with minimized mast bending. This makes it convenient to stress all cables from within one chamber at each mast head without the need to move the jacks over long distances.

With that method there was no need any more to stress the cables at their anchorages along the deck, which meant these bottom anchorages could be really simple and made from completely prefabricated steel elements. Their vertical plate provides the cable anchorages; ribs or teeth welded to it introduce the horizontal components of the cable forces into the deck; the horizontal bottom plate provides the vertical support; finally a transversal prestress is needed. This anchor detail was easily expanded to serve for the back-stays as well, including pendulum eye-bars to hold down the vertical and providing for horizontal movements simultaneously.

The Evripos Bridge clearly demonstrates the possible simplicity of cable stayed bridges and I do not want to give up the hope that such a design will some day return to the developing world where it basically came from.

An engineer working in developing countries with their overpopulated cities and their agonizing poverty resulting in human indignity, coming from a saturated part of the world where it is fortunately more and more understood that we cannot continue depriving this earth and that we are all living on the same globe, will try to understand the reason for such miserable conditions. So, to give an example, I sketched from World Bank statistics a graph and illustrations (Fig. 7.7), which describe some simple but dramatic interrelations:

- the lower the standard of living in a country (expressed by the gross national product per capita), the higher is its population increase,
- the energy consumption per capita in a country is proportional to its standard of living,
- therefore if a country wants to reduce its population increase, it has to increase its energy supply;
- to understand that please contemplate the following paradox: the larger the share of agricultural production in the total gross national product of a country, the more people in that country suffer from hunger!
- if the developing countries with their immense population (increase) satisfy their future energy demand from fossil or nuclear sources, these resources will not last nor will the environment;
- on the other hand the poor countries mostly are those with high solar irradiance. They would need to utilize only a small percentage of their desert areas to supply the whole world with solar electricity;
- therefore the developing world should be stimulated and supported to enter the new age of solar energy utilization which is the obvious answer to overpopulation, job procurement and environmental protection.

The question remains whether the technologies for solar utilization are sufficiently advanced to permit an affordable and long lasting electricity generation including the storage and transport of that energy to the Northern industrialized countries. I tried to give the answer 'yes, they are' already in my gold medal address in 1991 [1] and I have shown there, that small and large size solar power plants are a new and rewarding activity for structural engineers. Therefore it may be suffi-

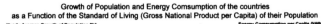

Growth of Population and Energy Comsumption of the countries
as a Function of the Standard of Living (Gross National Product per Capita) of their Population

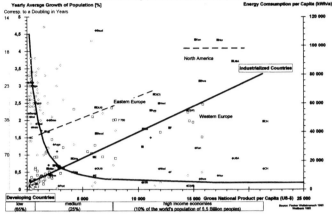

(a)

24.03.1994 BSP-ENEE.XLC

Stuttgart

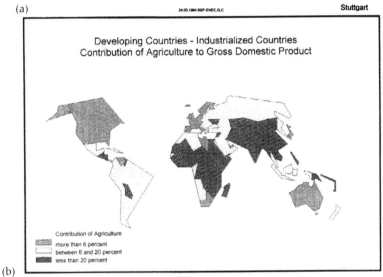

(b)

Developing Countries - Industrialized Countries
Contribution of Agriculture to Gross Domestic Product

Contribution of Agriculture
- more than 6 percent
- between 6 and 20 percent
- less than 20 percent

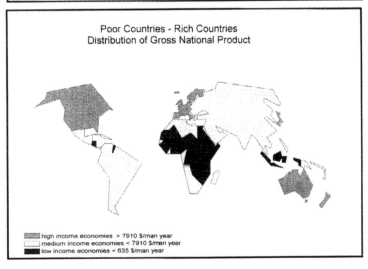

(c)

Poor Countries - Rich Countries
Distribution of Gross National Product

- high income economies > 7910 $/man year
- medium income economies < 7910 $/man year
- low income economies < 635 $/man year

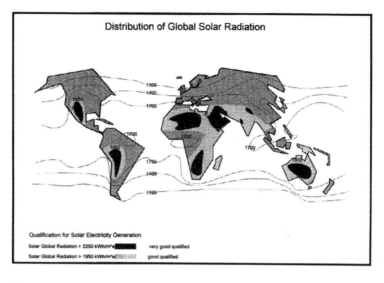

(d

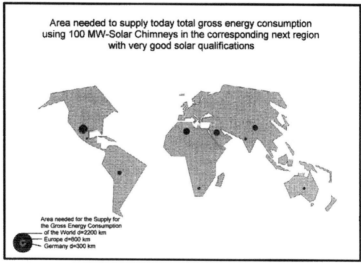

(e

Figure 7.7 (a) Standard of living, population increase and energy consumption for industrialized and developing countries; (b) and (c) agricultural countries suffer from hunger; (d) and (e) the poor countries have ample sun.

cient here to show only one typical small scale facility, a Dish/ Stirling system (Fig. 7.8a), where a metal membrane concave concentrator heats a gas in its focus which drives a Stirling-engine coupled to a generator (with 7.5m diameter of the dish the output is 9kW) and one large scale power plant. In the solar chimney (Fig. 7.8b) under a large glass roof solar radia-

(a)

(b)

Figure 7.8 Examples for small and large scale solar energy utilization:
(a) a Dish/Stirling system; (b) a solar chimney plant. © Prolab GmbH.

tion warms the air which flows to the base of a vertical tube
where it creates an upwind and drives turbines (with 3.600m
diameter of the roof and a 950m high chimney the output is
100MW).

What is still missing is a large scale demonstration; there
exist only small though successful prototypes. Unfortunately
our society seems to be unable to really care for the poor in the
developing world and for the environment unless the catastro-

phe has really happened, when in this case it may be too late. We engineers should leave our technological ivory towers and interfere with politics and tell those who have the power how and where to use it!

Reference

[1] Schlaich, J. (1991) Gold medal address. *Structural Engineer*, **69**, 10.

8
Research and development

Michael R. Horne

Professor Horne graduated in engineering at Cambridge University in 1941, returning to Cambridge in 1945 to work with the late Professor J.F. Baker on the development of the plastic theory of steel structures. Following appointments there as an Assistant Director of Research and Lecturer, he was appointed to the Chair of Civil Engineering in the University of Manchester in 1960, retiring in 1983. His main concerns in research have been on the stability problems of steel structures in the elastic–plastic range. He served as the president of the Institution in the year 1980–81.

Of all the interacting activities of the Institution, that of encouraging, guiding, and promoting research, of helping to communicate and publish its findings, and of aiding its application through the professional skills and achievements of its members, most clearly expresses the intertwining of our roles both as a learned society and as a chartered body of professionals. The Institution recognizes the centrality of its function in harmonizing these activities and bringing into effective partnership the individuals concerned in various ways. Research workers in both governmental and industrial research organizations, and within firms and in universities are able to achieve professional recognition and full membership, thereby making themselves the more readily available and motivated to take part in the many activities of the Institution in which experience and knowledge of research is essential.

Within this context, various important activities of the Institution come readily to mind. The contents of our journal, *The Structural Engineer*, express our role as a learned society in communicating the results of research, allowing valuable opportunity for discussion and criticism both by other research workers and by engineers engaged in design and construction. From time to time, as the need has been presented to the Institution, specially convened committees or study groups

Structural Engineering: History and development, Edited by R.J.W. Milne. Published in 1997 by E & FN Spon, London. ISBN 0 419 20170 X.

with memberships covering appropriate ranges of experience in consulting, contracting and research have depended on the devoted voluntarily given efforts of their members to allow the gathering and assessment of the state of knowledge on a subject. The Institution has then published reports on the resulting recommendations. The Institution is relied upon to appoint suitable members for service on Codes of Practice Committees, where again the results of research can be brought to bear on practice. I would not feel myself capable of providing any sort of useful assessment or summary of these activities over the past 60 years, and to attempt any overall review of 'research and development' in structural engineering generally over such a period would be absurdly pretentious.

My own career in research having been concerned over many years with investigations into the ultimate strength of steel building frames, it is in this field that thoughts most readily come to me on how research is motivated, planned and carried out and of how progress is made, sometimes ploddingly, sometimes with greater inspiration and quicker success. An article in *The Structural Engineer* for 1929 describes the setting up, by the Department of Scientific and Industrial Research (DSIR), of the Steel Structures Research Committee (SSRC) with the task of applying 'modern theory' to the design of multi-storey steel building structures [1]. This was the result of a proposal by the British Steelwork Association (BSA), who then undertook to share equally with the DSIR the cost of financing the planned five year investigation. While there had not appeared to be any manifest reason to doubt the basic safety of buildings designed to the then widely used regulations of the London County Council (later used largely unchanged as the basis for the first BS449 (1932)), the apparent illogicality of design procedures in which beams were treated as simply supported while columns, making use of the 'effective length' concept, were assumed to derive support from partial fixity from bolted or rivetted beam to column connections, had led to the belief that more rational procedures could lead to improved, more economical designs.

The three reports of the Steel Structures Research Committee [2], published over the years 1931–36, contain a record of research, remarkable for its time both in its extent and its thoroughness, that was carried out under the auspices of the Committee on all aspects of the behaviour of multi-storey steel frame buildings as commonly designed at that time. Included, in addition to appreciable theoretical investigations, were ex-

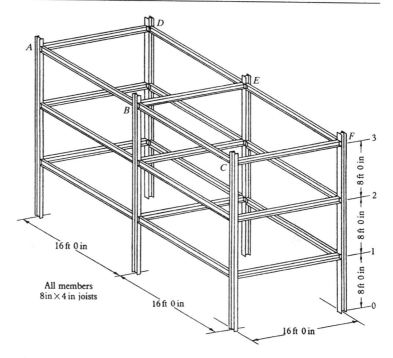

Figure 8.1 Experimental test frame erected at the Building Research Station.

tensive experimental investigations into the moment/rotation characteristics of beam to column connections carried out by Professor Batho at Birmingham, and the measurement of strains (and hence deduced stress changes) in loading tests on the multi-storey steel frames of a museum building, a hotel, an office building and a block of flats, and on an experimental three-storey, two-bay frame (Fig. 8.1) erected at the Building Research Station.

There is no doubt that these most extensive, originally conceived programmes of research set a pattern which continues to this day. In fact, 60 years later, we are still engaged in the testing of large, specially constructed multi-storey steel frames, the latest and most impressive manifestation of which is the eight-storey, five bays by three bays frame (Fig. 8.2) in course of preparation for testing at the BRS Cardington Laboratory, as described in a recent article in *The Structural Engineer* [3]. The highly ambitious project (of which this will be the first fruits) aims at investigations, not only into the static structural interactions between steel beams and columns and the

Figure 8.2 Eight-storey experimental steel framed building at BRE Cardington laboratory. (Reproduced by kind permission of the Building Research Establishment, Garston, Watford.)

consequential conclusions on how improved methods of design may be arrived at, but also into the behaviour of a sufficiently realistic cladded steel frame structure in relation to serviceability criteria, to vibrational and explosive loading and to structural survival under thermal loading due to fire. Further investigations over a 10 year period are envisaged on reinforced concrete, timber and masonry buildings.

Our topic is research and development over the past 60 years. How, may we ask, has the way of research changed over that time? Changes one would expect, and there have indeed been many. But here, on the basis of the strength design of multi-storey steel frames we seem to be faced with understanding how both the topic and the method seem to have remained entirely the same. What does this tell us about the nature of and the motivation for structural engineering research? Or how would we, for that matter, meet the comment of the sceptic who would say that, despite so much effort over many years, we do not seem to have learned very much at all?

Numerous articles and discussions on the purposes and motivations for structural research and comments on its organization and achievements have appeared regularly in the pages of the Journal. Notably, Professor Sir Alfred Pugsley in his Presidential Address in 1957, 'The way of research' [4], expounded how research found its motivation and purpose through problems in the design office, in the development of Codes of Practice and from problems arising from the use of new materials. Professor Pugsley gave examples of how demands for research had arisen urgently in the past from pioneering projects, as in the design of large bridges and aircraft structures, and also discussed the particular role of those working in universities in their pursuit especially of fundamental research. Dr R.E. Rowe, in his Presidential Address in 1983, 'To research communicate, and codify – well!' [5], concerned himself particularly with the planning and guidance of research motivated directly by problems arising in design, in its communication to practising engineers and the suitable incorporation of research results into Codes of Practice. While asserting the desirability of leaving the management and execution of research 'to the experts, or professionals, in that field, be they associated with the Government research establishments, the research associations, the universities or industrial research laboratories', he emphasized the importance of the Institution's role in bringing intimately to bear the wisdom of practising structural engineers in both guiding the definition of research needs, in aiding the influence of research on design and construction, and in assisting in its suitable incorporation in Codes and Standards.

There has been evidence in abundance throughout the last 60 years, through addresses, papers, reports and discussions, of the awareness of the Institution of the importance of all these aspects. To mention but one well reported discussion, one may refer to that on 'Structural engineering research in Britain' which took place in 1966 [6]. Speaking then, another past president, Brian Scruby, delineated four categories into which research could conveniently be divided, and elaborated on where he saw the main responsibilities desirably lay in initiating the research which should be undertaken. 'Basic' research, into the properties of new materials such as higher strength steels or concrete or of the structural properties of materials not currently used in structural engineering, he saw as essentially the role of the research worker. In 'theoretical research into stress analysis', involving ways of improving our theoretical understanding of the laws of structural behaviour and methods of computation directed towards the achievement of stable and economical structures, he saw the necessity of not leaving the setting of aims and priorities entirely to the research worker. 'Practical research', involving investigation into the problems of the designer and the constructor in their day to day work, necessarily had to be initiated by the practising engineer, while at the same time being the most difficult category in which adequately to bring about the necessary degree of cooperation between practising engineers and research workers. Similar considerations obtained in the final category of 'development research', involving ventures into new forms of construction as had been exemplified by the development of industrialized building and of composite construction in steel and concrete.

It would be tempting to try to imagine the ideal progression there could have been, through such categories of research, of just such an investigation as that into the behaviour and design of multi-storey steel frames, if only we but had the wisdom to plan it and execute it. Perhaps we would not then be finding it necessary still to be carrying out such an expensive testing programme on still more full scale frames! But the problem is that, although it can help in our thinking to set ideal aims for the way in which research should be categorized and organized, the various types of activity, problems and innovations are all inextricably mixed and interacting the one with the other.

The setting up of the Steel Structures Research Committee was, as we have seen, very much a joint exercise between research workers and the steel construction industry, within which virtually all current design of multi-storey frames was carried out. The thoughts of consultants were, for practical purposes, expressed in the building regulations of the times. Research and testing, both on joints and on full scale frames, was restricted to joints and frames as designed in the then current practice. In the end, this practically conceived research did not lead to any improved economy in design procedures. There is no guarantee at all that 'practically conceived' research, in the 'practical research' category, will necessarily be 'successful'. Unfortunately, things are not so simple! Professor J.F. Baker who had been the experimental officer for the SSRC investigations, came to the conclusion that a reason for this failure was the restriction to the use of limiting elastic stress criteria as the basis for the drawing up of the design recommendations. This led to his pursuit of the investigations leading, after the Second World War, to his advocacy of the so-called 'plastic methods' of analysis and design with which his name is intimately associated, and formed my own introduction to the areas of research which have been my main concern.

Here it may be of interest to note that, following the first award of the Institution's Gold Medal to Henry Adams in 1922, it was not until 31 years later that the Institution awarded its second Gold Medal to John Fleetwood Baker. I (and no doubt my fellow authors for this commemorative volume) may be grateful that Professor Baker thus provided the opportunity for the Institution to revive the custom of making the award!

Looking back on the history of this research in which Professor Baker was involved provides an example of many of the features of that type of engineering research and development which involves the introduction of newly applied concepts, materials and procedures as opposed to the more routine testing and data accumulation related to established practice. There has to be some initial, physical or theoretical concept that will inspire the research, sufficiently simplified to fire the imagination. The goal of a defined material creation that directly inspires the engineer in practice is not available. For that early work on 'plastic theory', that simple concept was found in the easily visualized ideas on 'plastic hinge mechanisms', which J.F. Baker described in a paper to the Institution in 1949 [7]. At that time, as a very young lecturer to a post-graduate course in the Cambridge University Engineering Department under Professor Baker, I was daunted to hear that Sir Richard Southwell, at that time retired and living in Cambridge, had asked whether he might sit in on my lectures on plastic theory. This he did two or three times, but then gave up, expressing his thanks but saying to Professor Baker that 'it was really all very simple'. At the time, I was a little chagrined, feeling that the great man was expressing disdain. But I may have been wrong and perhaps I should have accepted it (intended or not) as a compliment!

It is also a feature of innovative research that, while the initial inspiration is stimulated by the simple idea, it does not stay that way. The more sophisticated perception or prejudice, take it as you will, of the practising engineer usually finds expression, and has to be met. It is usual to find that at least some of the reservations of practising engineers are justified. In this case, following many small scale laboratory tests, more convincing tests up to collapse on full scale portal frames were carried out, using dead loading rather than jacks to make the loading more representative of real structural loading (Fig. 8.3). The course of this research, while providing evidence of the basic soundness of the fundamental concepts of plastic theory, demonstrated some of its limitations, showing that both local instability in the form of lateral–torsional buckling of I-section members and overall frame instability needed to be considered.

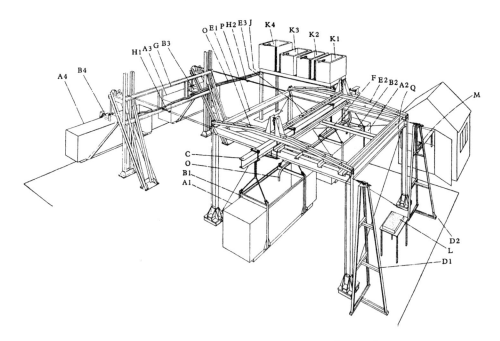

Figure 8.3 Investigations into the plastic theory of structures. Experimental portal frame erected at the BWRA laboratories, Abington, Cambridgeshire.

A main feature of structural research over the past 60 years has been an increasing appreciation of the concept of 'limit state design', despite the fact that the term has been variously interpreted and commonly misunderstood. As interpreted in the various 'Limit State Codes' which have been developed, it refers mainly to the two main concepts of the limiting states of strength and serviceability, the former being interpreted primarily in terms of limit states of collapse at factored loads, the latter in terms of limiting deflections at working loads. For the purposes of design, these terms need to be reduced to limiting requirements calculable by reference to analytical procedures based on the physical properties of the materials of construction, including yield or ultimate failure under simple or combined stresses, the conditions for brittle fracture or failure due to fatigue, the value of the modulus of elasticity or, in more general terms, the entire nature of the stress-strain relations. Unfortunately, these concepts can become complicated and difficult of calculation, and idealizations and approximations become essential. It is here that trouble has arisen in a running controversy between those who are prepared fully to accept the new explicitly 'Limit State Codes', and those who wish still to use only former 'Safe Stress Codes'. In fact, the controversy is a false one, since all are in fact 'limit state' in the true sense – the only difference is in the nature of the 'idealizations and approximations' involved. The long success of the 'safe' stress concept lies in the fact that, of all the single calculable responses within a structure, the nominal theoretical maximum elastic stress is the one most capable of conservatively encompassing and controlling the strength of the structure in its response to loading. What has worried research workers in their attempts to improve our understanding of structural behaviour is the fact that, the more one investigates the actual stresses induced, the more one becomes aware that these, as opposed to the nominal stresses calculated when using Safe Stress Codes, have no particularly consistent relationship to real strength. The task of the research worker then becomes one of finding a more rational basis of relating real behaviour to theoretically calculable quantities, and of assisting their incorporation in Codes of Practice.

An immediate consequence of introducing the concept of plastic hinge theory was the realization that it provided an immediate explanation of why residual stresses due to the rolling or welding of steel structural members, or to the fabrication of redundant steel frames, had no effect at all on failure loads unless brittle fracture is involved, except for possible secondary effects where elastic stability is of major importance. Thus when using 'safe stress' procedures, one may neglect internal self-balancing residual stress states even when these involve stresses up to the yield value.

One of the main incentives for the pursuit of plastic theory was disillusion with safe stress criteria as applied to the design of continuous columns. Despite much effort expended on maximum stress column design procedures recommended in the Final Report of the Steel Structures Research Committee, the design method led to excessively heavy columns as compared with BS449. Since there was no reason to believe that the use of BS449 had led to the construction of unsafe structures, the conclusion had to be that the safe stress criterion was an inadequate basis for the design of continuous columns. The intuitive equivalent length procedures used in BS449 for dealing with the stability of continuous columns was evidently more successful than an apparently more rationally complete safe stress procedure, thus indicating the necessity for a completely new look at the stability problem.

One interesting new approach was due to Merchant, who in a paper in the Journal in 1954 [8] proposed that failure load factor λ_F might be expressed as some function of various load factors calculated using idealized assumptions. In particular Merchant suggested a generalized formula for λ_F, subsequently called the Rankine–Merchant formula and found to have some theoretically justified basis [9], as providing an approximate lower bound to the failure load in the form

$$\lambda_F = \frac{\lambda_P \lambda_C}{\left(\lambda_P - \lambda_C\right)}$$

Here λ_P is the simple plastic collapse load and λ_C is the elastic critical load. One may think of λ_P as the failure load of a frame with material of infinitely high elastic modulus while λ_C is the failure load of a frame when the yield stress is infinitely high. In this way, the influences of the two main physical properties that influence the ultimate load of a mild steel structure are explicitly introduced, helping towards a clearer understanding in simple terms of the nature of their interacting relationship. Merchant's original, admittedly tentative, justification was based on a series of theoretically calculated elastic–plastic failure loads of one and two storey rigid frames [8] (see Fig. 8.4) which demonstrated a clear tendency for a high proportion of

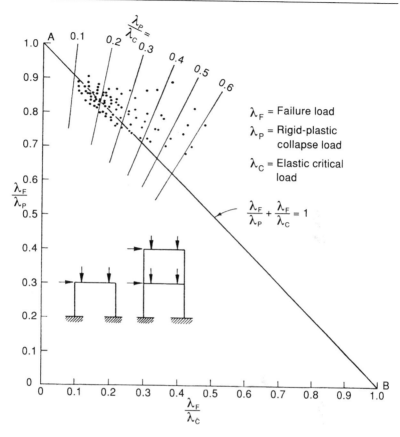

Figure 8.4 Rankine–Merchant load compared with theoretical failure loads of one-storey and two-storey frames. ••

the theoretical loads to crowd round the Rankine–Merchant line AB. Those lying well above were those with small or zero lateral loading, a feature again found to be capable of theoretical explanation [9].

The work pioneered by Merchant on the failure loads of frames provides an example of the way in which advances in structural understanding can follow from a combination of theoretical advances, intuitive insight and the use of computers to aid in the detailed exploration of the application of simplified theoretical models. A great deal of work of this nature has been carried out over many years at the Building Research Station, and most notably under the inspiration of the late Dr R.H. Wood. These days, the use of computerized procedures for analysing the behaviour of structures,

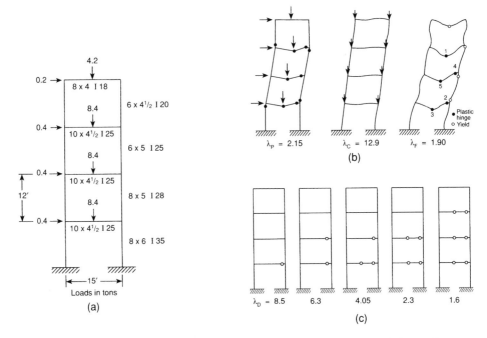

Figure 8.5 Theoretical investigation by Wood [10] of elastic-plastic concepts for the failure of multistorey frames: (a) details of frame; (b) rigid-plastic, elastic critical and elastic-plastic failure modes; (c) concept of deteriorated critical loads.

allowing completely for the whole physical behaviour of all the structural materials and components, has become commonplace, but this was not so for the analysis of the one and two storey frames quoted by Merchant for which the work was certainly quite onerous. Wood himself made an important contribution to theoretical understanding by his work on the progressive deterioration of the critical loadfactor λ_D due to the successive formation of plastic hinges [10] (see Fig. 8.5), but most interestingly the computational work which he used to illustrate his concepts was carried out, not by digital computation, but by the use of an analogue computational mechanism in the form of a differential analyser, a *tour de force* one might say, if ever there were one!

With the passage of 60 years since the investigations of the Steel Structures Research Committee, can one say that our Codes adequately and clearly profess, openly to the user, the basis and justification for the clauses they contain? While undoubtedly the fundamental work of the type I have described has contributed greatly in the development of our Codes, it is doubtful whether the average user is greatly the wiser regarding the deeper structural significance of the detailed requirements. The demand is for Codes that are essentially as tightly prescriptive and comprehensive as possible, yet sufficiently simplified despite any complexities of structural behaviour. While one side of me readily accepts this as probably inevitable, I cannot help wondering whether we are sometimes content to allow this requirement to set our sights too low in meeting the intellectual challenges of our profession.

An illustration of the difficulties facing any higher hopes may be provided by the fate of a work of very considerable achievement on the part of R.H. Wood, who in 1973 produced what to me was a quite brilliant paper [11] in which he attempted to bring together all that had been learned about the complex behaviour of columns in multi-storey buildings from theoretical, computational and experimental investigations. He proposed a method of design which aimed at taking into account the interactions between continuous columns and the attached beams framing into both column axes, allowing for the elastic–plastic behaviour of all the members including where significant the phenomenon of torsional–elastic failure of I-section columns. Necessarily this would have presumed some willingness on the part of the user to encompass some understanding of the quite complex structural concepts involved. I remember feeling at the symposium at which Wood's proposals were expounded that, while the audience were prepared to acknowledge the ingenuity with which he had attempted to deal with a complex problem, there would be little hope of any acceptance of his proposals in practice. The importance thus has to be incorporated by research workers, accepting the limitations as well as the possibilities of their roles in helping to apply their work in Codes of Practice, for which purpose the aim has to be to achieve a safe, reasonably easy to understand and economical method of design without necessarily aiming at the ultimate in complete rationality and economy. In helping towards this aim, the Institution deserves considerable credit for having successfully encouraged an appreciable number of able research workers who, while accepting those limitations, have laboured to make major contributions to the work of Code of Practice Committees in incorporating the results both of their own research and that of others in the field.

Because of my own interests in that field, I have drawn my examples of the way in which research is inspired and concepts inspired and developed from the field of structural steel building frames. It is an area in which the inspiration for research has mainly arisen from dissatisfaction with the rationality of existing design procedures rather than from structural catastrophes arising from their use. A similar case has arisen in the field of structural concrete in relation to the problem of shear failure in beams. This is a subject which, despite continuing attention for a great many years, still seems to defy any entirely successful theoretical solution. P.E. Regan has, in a recent paper in the Journal [12], given a review of some of the many theoretical solutions that have been proposed. By calling his paper 'Research on shear: a benefit to humanity or a waste of time?' he reflects what could be a reaction to the enormous number of papers that have been written on the subject. Regan lists 61 references, which is doubtless only a small proportion of the total number. As opposed to the relative simplicity of the elastic–plastic model for mild steel, it is obviously much more difficult to produce an adequate, all-embracing mathematical model for the multi-phase material which constitutes concrete itself, to which has to be added the multiplicity of reinforcing arrangements and steel/concrete interactions which need to be taken into account. A great difficulty here is that, as Regan points out, a model that gives good correlation between test and theory for one range of concrete mix and detailing may not be at all satisfactory for another.

Research investigations can be urgently triggered by the experience of catastrophic failures. Examples of this have been in the incidence of brittle fracture in welded structures, of the failure of box girder bridges and in the saga of high alumina cement. With hindsight, such failures can all too often be seen as revealing lack of foresight of the effects of known facts rather than the sudden appearance of phenomena of which there has been no previous knowledge. It is interesting to note that as far back as 1930, there appeared in the Journal a draft Specification for High Alumina Cement [13], covering test requirements for 'fineness, chemical composition, compression strength, setting time and soundness'. This was the result of two years work by the 'Sectional Committee on Concrete' of the Institution, and was the subject of the discussion which followed immediately after the address of the then newly installed President R.H. Harry Stanger. In reply to a question, the President said that he believed it to be the first Specification for HAC 'not only here but in the world'. The serious deterioration in the strength of structures containing HAC concrete that could occur with time had by then already been observed, and advice on its use was contained in a DSIR report in 1934 [14] and in a report of our Institution in 1937 [15]. Its advantage in allowing rapid turnround in the manufacture of precast units nevertheless led, after the Second World War, to its widespread use, and while warnings were issued advising caution on required mix proportions and the range of circumstances in which its use was inadvisable, it took a long history of field experience of revealed structural failures to bring about a sufficient understanding and acceptance of those conditions and limitations [16,17].

It is attractive to have a vision of a well-planned and ordered progression through a rational sequence of basic research on to the strategic development of increasingly ambitious practical applications. The experience in practice is a great deal more complex and hesitant. Research and development go hand in hand, and the achievement of their effective interplay has to be for the research worker his main inspiration, and for the practising engineer an essential component of the innovations he seeks to introduce.

References

[1] Baker, J.F. (1933) The design of steel structures, *The Structural Engineer*, **7**, 333.

[2] Steel Structures Research Committee, First Report (1931), Second Report (1934) and Final (1936) Report. HMSO.

[3] Armer, G.S.T. and Moore, D.B. (1994) Full-scale testing on complete multistorey structures, *The Structural Engineer*, **72**, 30.

[4] Pugsley, A.G. (1957) The way of research, *The Structural Engineer*, **35**, 403.

[5] Rowe, R.E. (1983) To research, communicate and codify – well!, *The Structural Engineer*, **61A**, 371.

[6] Structural engineering research in Britain, Report of a debate held at an Ordinary Meeting of the Institution on Thursday 13 January 1966. *The Structural Engineer*, **44**, 1966, 215.

[7] Baker, J.F. (1949) The design of steel frames, *The Structural Engineer*, **27**, 397.

[8] Merchant, W., Rashid, C.A., Bolton, C.A. and Salem, A. (1958) *The behaviour of unclad frames*, Proceedings 50th Anniversary Conference, Institution of Structural Engineers.

[9] Horne, M.R. (1963) Elastic-plastic failure loads of plane frames, *Proc. Roy. Soc. A*, **274**, 343.

[10] Wood, R.H. (1957–8) The stability of tall buildings, *Proc. Inst. Civ. Engrs.*, **11**, 69.

[11] Wood, R.H. (1973) *A New Approach to Column Design*, HMSO.

[12] Regan, P.E. (1993) Research on shear: a benefit to humanity or a waste of time?, *The Structural Engineer*, **71**, 337.

[13] Tentative specification for high alumina cements, *The Structural Engineer*, **8**, 1930, 369.

[14] *Report of the Reinforced Concrete Structures Committee of the Building Research Board with Recommendations for a Code of Practice*, DSIR, HMSO, 1934.

[15] *The use of high alumina cement in structural engineering* (Report no. 21) Institution of Structural Engineers, 1937.

[16] Safier, A.S. (1980) High alumina cement concrete – appraisals, the problems, and some findings, *The Structural Engineer*, **58A**, 381.

[17] Bate, S.C.C. (1980) High alumina cement concrete – an assessment from laboratory and field studies, *The Structural Engineer*, **58A**, 388.

9

Innovation in structural engineering – challenges for the future

Alan Davenport

Alan Davenport's career has been blown by the wind. His research and advice on wind loading have helped shape structures world-wide. These have included some of the tallest – the World Trade Center in New York, the CN Tower in Toronto, the Sears Building in Chicago, the Hongkong Bank Building and the Bank of China in Hong Kong, – as well as some of the longest – the Tsing Ma Bridge in Hong Kong, the Normandy Bridge in France and the Storebaelt Bridge in Denmark. Dr Davenport has been the founding director of the Boundary Layer Wind Tunnel Laboratory at the University of Western Ontario, Canada for the past 35 years. His research on wind turbulence and the dynamic response of structures to wind has pioneered new approaches to treating wind action and has been active in the development of new wind load codes used around the world. Dr Davenport's other contributions include the climate and meteorology of windstorms, structural dynamics, statistical methods of seismic zoning, truck loading of bridges and structural safety. The recipient of a number of awards and honorary degrees, he is also a past president of the Canadian Academy of Engineering and has been named a foreign associate of both the Royal Academy of Engineering in the UK and the National Academy of Engineering in the USA.

Introduction   *'Innovation: A change in the established order; a new shoot on an old stem; (in Scot's law): the renewal of an obligation to pay'.*
Websters Dictionary [1]

Innovation has been and is a compelling factor in structural engineering which strongly affects its competitiveness, profitability and the value of its contribution to society. Innovation can relate to structural form, to function or scale; to materials, to methods of construction, or to analytical and design ap-

Structural Engineering: History and development, Edited by R.J.W. Milne. Published in 1997 by E & FN Spon, London. ISBN 0 419 20170 X.

proaches; to the nature of the regulations, or to the control of quality and last, but by no means least, to the education of the next generation of structural engineers.

This paper offers a few examples of innovation as a tribute to the innovative spirit of the Institution of Structural Engineers and to the structural engineers it has inspired through the past 60 years.

Innovation and tall towers

'Come,' they said, 'let us build ourselves a city and a tower with its top in the heavens, and make a name for ourselves . . .'

Genesis 11, 2–9

Tall towers have always been the symbol of structural engineering achievement and innovation. They have served a multitude of purposes. The early ziggurats expressed power and the ambition to approach closer to heaven (Fig. 9.1) and in defence they provided a commanding position for viewing and firing at the enemy. In the Middle East the towers scoop wind to cool the buildings through evaporation. In medieval Europe they advertise the importance of the owners. In radio and microwave communications they provide longer 'line of sight' range. Japanese construction companies vie for the 1000m towers for a continent where the population is rising and land becoming scarcer. In modern cities they provide a striking skyline and a perspective of a city's urban hustle which by day is peaceful and at night, fairyland (Fig. 9.2).

Innovation and roofs

'The singing masons building roofs of gold . . .'

William Shakespeare
Henry V, 1. Chorus

Figures 9.3–9.11 demonstrate the variety of roof shapes developed over the years in response to different design demands. Some of these demands arose from functional requirements, others from the aesthetic appeal. In the absence of codes of practice we may assume that a kind of Darwinian evolution is at work in which, through careful observation of performance, the more cost effective, functional and satisfying forms would be selected over those which were less successful.

The thatched roof houses of the Shetland Islands (Fig. 9.3) offer a low profile and porous roofing material that reduces the differences between external and internal wind pressure and consequently the loads and rain penetration. The advantage of lightweight materials and structural integrity become apparent when mobility is everything (Fig. 9.4)! Contrasting gable roofs

Figure 9.1 An early ziggurat.

in Denmark (Fig. 9.5) suggest conflicting design considerations, with the use of thatch and tile, eaves and no eaves (with the structural vulnerability of the eaves on the one hand and the protection they afford on the other). The traditional tent (Fig. 9.6), though simple in construction, still offers protection

Figure 9.2 A telecommunications tower.

against the sun. A high-tech answer to covering the largest roof area in the world was achieved by the use of Teflon coated fibreglass (Fig. 9.7).

Innovation: the hip versus the gable roof

'There was a most ingenious Architect who had contrived a new Method for building Houses, by beginning at the Roof, and working downwards to the Foundation; which he justified to me by the like Practice of those two prudent Insects the Bee and the Spider.'

Swift, Gullivers Travels

Figure 9.3
Thatched roof
houses in the
Shetland
Islands.

Figure 9.4 The
light weight of
thatched roofs is
an advantage.

9.3

9.4

Figure 9.5
Gable roofs in
Denmark.

Figure 9.6
Traditional
nomadic arab
tents.

9.5

9.6

Figure 9.7 The
Haj terminal in
Jeddah,
designed by
Fazlur Khan.

9.7

Figure 9.8
Louvre in Paris,
entrance
designed by
I. M. Pei.

Figure 9.9
Lotus motif of
the Bahai
temple, New
Delhi.

9.8

9.9

Figure 9.10
Roof skyline in
Prague.

Figure 9.11
Shogun castle.

9.10

9.11

Figure 9.12 A hip roof after a hurricane.

Figure 9.13 Hurricane damage to a gable roof.

Post-disaster visits to the sites of windstorm disasters have indicated that certain forms of construction have superior survival characteristics. One of these is the hip roof. The familiar chamfered end of this roof type and the squared off end of the gable roof are illustrated in Figs 9.12 and 9.13. Both photographs were taken following hurricanes. The hip roof is traditional in many Caribbean countries, and elsewhere; the gable

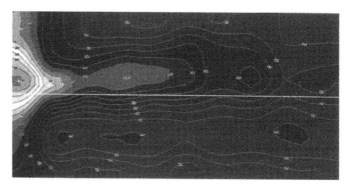

Figure 9.14 Pressure distribution on a gable roof for quartering winds ($C_{pi} = +0.25$).

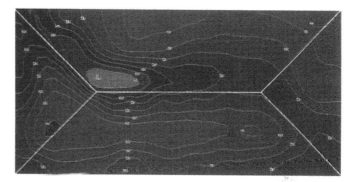

Figure 9.15 Pressure distribution on a hip roof for quartering winds ($C_{pi} = +0.25$).

roof is the modern standard, more convenient to build due to the availability of mass produced roof trusses.

To investigate this apparent advantage of the hip roof, tests in a boundary layer wind tunnel were undertaken comparing the pressure distributions on the two roofs. The results are illustrated in Figs 9.14 and 9.15. These clearly show that the uplift suctions on the hip roof are highest at the peak of the roof where the members and surfaces intersect and the roof is considerably strengthened. On the other hand, on the gable ended roof, not only are the suctions higher but the maximum suctions occur along the entire length of the end gable. The end trusses must carry the brunt of the wind force particularly if the roof is cantilevered to provide for eaves. On all these counts the hip roof is better and, for the same amount of material, can be considerably stronger. Innovation may have been running in reverse.

Innovation and damping

'It will not be long before structural engineers are faced by an urgent problem: how to endow their structures with artificial damping'.

R.V. Southwell, 1949 [2]

In recent years structures have become lighter and more flexible due to the increased strength of most structural materials and the disappearance, for example, of heavy masonry infill and cladding in building. This has rendered structural damping less and at the same time left structures more prone to resonant vibration. The dominant structural loads of wind and earthquake as well as traffic, have raised the importance of the dynamic response and rendered it the most significant.

Structural damping as a design factor was introduced into The World Trade Center over 25 years ago as a key component in its armour against the dynamic excitation by wind. Over 10 000 visco elastic elements were introduced between the bottom chords of the floor trusses and the exterior structural columns. These elements were stretched or compressed during the sway of the building and thus absorbed energy. Damping was subsequently included in a number of other structures and investigated as part of the wind response – these included the Centrepoint Tower (Sydney, Australia – tuned additional mass damper), The CN Tower (Toronto – tuned mass damper), the Citicorp Building (New York – servo-assisted tuned damper), and the John Hancock Building (Boston – servo-assisted tuned damper). It has also been incorporated in longspan suspension bridges such as the Bronx-Whitestone Bridge (New York – friction damper).

In these, additional structural damping was required to nail down more accurately the available damping, to offset any negative aerodynamic damping and the tendency for aerodynamic instability and large amplitude resonant sway.

After a slow start, damping has now become a more important part of the weaponry in controlling the response of structures to wind and earthquake.

Innovation and codes of practice

Evolutionary improvements, exemplified by hip roofs and structural damping, can easily be blocked by codification – nowhere, it would appear, do codes point explicitly to the improved performance of hip roofs. While the outward purpose of codes is to distil contemporary wisdom and experience and protect the public, in the way they work there can be clear disadvantages. They are often ponderously slow to change; they deflect some responsibility for safety from the designer, and in many circumstances are mute on innovative measures.

The replacement of prescriptive type codes by performance based codes is generally a step forward.

Innovation and natural hazards

'Blow, winds, and crack your cheeks! rage! blow!
Your cataracts and hurricanoes, spout
Till you have drenched our steeples, drowned the cocks!'
William Shakespeare
King Lear, III.ii

Natural hazards such as windstorm and earthquake are the ultimate test of structures' load carrying capacity. Currently there is concern over the increasing costs of natural catastrophes which have, over the past decade, tripled. On an annualized basis, windstorm is now the most costly of these hazards; earthquake the next most costly. These increased costs have made heavy inroads on the reserves of the insurance industry and their ability to pay. By the same token the availability of insurance is becoming more difficult in those areas most vulnerable to these hazards.

There are indications that the responsibility for safe construction and improvements in quality needs to be spread more widely through the construction industry to include other stakeholders in construction. Codes can only influence the design of new construction. The involvement of the investors and users is needed to develop awareness of the risks and the initiative to improve the quality of construction for buildings new and old.

Innovation and the marketplace

Like most creative activity, structural engineering responds to the marketplace. In the context of manufactured products Akio Morita, president of the Sony Corporation has defined the links between scientific discovery, innovation and the marketplace in his remarkable paper '$S \neq T \neq I$' – 'Science alone does not equal technology; technology alone does not equal innovation' [3].

In this paper Dr Morita describes the differences between scientific discovery (of new natural phenomena materials and principles), invention (of a new process, device or structural form or function), and the innovation required to develop an invention into a viable, marketable and usable product. He stresses the need for engineers to become familiar with and creative in all of these areas.

Innovation and the education of the structural engineer

'There seems to be something in the work of the engineer which suppresses talk, even useful talk. This is very well in a way but can be carried too far. Engineers ought not to hide their light under a bushel and expect the world to reward them for their silent work's sake. The world is too busy . . . to study engineering and would perhaps take more interest in engineers if they were to take the trouble to explain things.'

John Galbraith, 1909 [4]

The education of the structural engineer has been under revision for centuries. Earlier education included apprentice-ship training (in the English speaking world), military academies, Ecoles des Ponts et Chaussées, Grands Ecoles, and universities.

Today the engineer that society seems to need is:

- A technical expert;
- a decision maker;
- a leader;
- a communicator;
- a financial expert;
- a risk assessor;
- an environmental expert; and
- a politician (perhaps).

What society often gets is just

- a good theoretical analyst;
- a conventional designer.

For some engineering schools this assessment may be ungenerous; but in most instances there seems to be a need for introducing students to a kind of problem solving which brings to life the social, economic, commercial and political context of real life problems as opposed to the 'serial learning' derived from the systematic progression through engineering texts. This lies outside the bounds of the normal design exercise. How can this be done within the normal constraints?

An approach which has been tried extensively in business schools, but rarely in engineering schools, is the so called 'case study' approach. Here one should comment that there are two models of case studies. The first is an historical reporting of engineering 'successes' (sometimes 'failures') which usually points to the 'right' way to do things. This type of approach can have great value but does not necessarily immerse the student in the thought processes of the participants. The second type of case study is commonly used in business schools for teaching problem analysis to business students.

This type of case usually consists of an actual situation which demands decisions to be made but not classified usually as either a success or failure. The 'case' consists of a brief describing the situation – the participants and their roles and responsibilities, the nature of the job they are doing, any background information needed. Students meet initially in groups to discuss 'the situation', identify what the problems are – technical, personal, political, etc. – and formulate decisions and the arguments in their support. After meeting in small groups to orient themselves they will meet in larger groups to compare the decisions proposed earlier in the smaller groups. The instructor is there primarily to animate the discussion not to 'instruct'. Various methods can be used to expand the experience such as written reporting.

A good case generally is 'a slice of life' which convinces the students they are participants in the situation at hand (hence the insistence on real situations). The subject matter is chosen for its content and difficulty. Cases may use collapses of structures to point out design errors, the responsibilities for uncovering such errors, what should be done to avoid their repetition. The issues may contain both technical as well as professional aspects.

The demands on the student are to:

- assess the situation in the case;
- formulate the problem;
- analyse the technical issues;
- propose solutions;
- identify obstacles;
- work with team;
- present results;
- take criticism;
- deal with social, economic, political and environmental concerns;
- carry out professional responsibilities.

The experience of students and instructors who have tried this approach is very positive; the students are interested in the introduction provided to the culture of the industry they will work in; and the sense of involvement.

Unfortunately the availability of such cases with a useful engineering content is not large. Many business school cases contain useful management issues. It is hoped that before long the collection of engineering case studies can be expanded.

Conclusion

'Because of its varied aspects, of its persistence in time, and of the scientific, technological, aesthetic, and social factors which influence it, construction may well be considered the most typical expression of the creativity of a people, and the most significant element in the development of its civilization.'

Pier Luigi Nervi [5]

Long live the spirit and example of the Institution of Structural Engineers.

References

[1] *Webster's New 20th Century Dictionary of the English Language*, 2nd edition, 1976, Collins World.

[2] Southwell, R.V. (1949) *Colston Research Symposium*, University of Bristol Press.

[3] Morita, A. (1992) Science alone does not equal technology; technology alone does not equal innovation. Paper given at the 1st UK Innovation Lecture at the Royal Society of London.

[4] Moriarty, Catherine (1989) *John Galbraith: Engineer and Educator*, p. 113: Presidential address to the Canadian Society of Civil Engineers, University of Toronto Press.

[5] Nervi, Pier Luigi (1955) *Structures*, McGraw Hill.

Index

Page numbers appearing in **bold** refer to figures.